# Fossil Invertebrates

PAUL D. TAYLOR AND DAVID N. LEWIS

Harvard University Press
Cambridge, Massachusetts
2005

First published by the Natural History Museum,
Cromwell Road, London SW7 5BD
© Natural History Museum, London, 2005

First North American edition published by Harvard University Press, 2005

ISBN 0-674-01972-5

Cataloging-in-Publication Data is available from the Library of Congress.

Edited by Rebecca Harman
Interior designed by Mercer Design
Reproduction and printing by Craft Print, Singapore

# Contents

# Preface

Fossils are the remains of ancient animals and plants. They provide us with the only tangible evidence of organisms that inhabited the Earth in prehistoric times, back to earliest indications of life some 3500 million years ago. Most fossils consist of long extinct species, sometimes almost identical to present-day species, but often so very different that a great deal of interpretation is needed before they can be understood. Deciphering the history of life as recorded by fossils extends beyond purely academic interest. Knowledge of changes in the Earth's biosphere over timespans of millions of years allows us to put into perspective the alarming environmental changes that are happening today. For example, we can compare the severity of extinctions caused by humankind's destruction of natural habitats – including tropical rain forests and coral reefs – with extinctions that happened in the geological past long before humans evolved. Fossils provide key information about the relative ages of sedimentary rocks, and about the environments in which they were formed, both matters of great importance to geologists in the search for natural resources such as oil and gas.

The study of fossils is called palaeontology. It is an interdisciplinary science occupying the ground between geology (the study of the Earth) and biology (the study of life). In addition to professional palaeontologists employed by museums, universities and geological companies and surveys around the world, fossils have a wide appeal to non-professionals. Most notably, they excite the interest and imagination of children and provide an easy entry into the wider world of science. Countless children and adults collect fossils as a hobby. Some are attracted principally by the sheer beauty of many fossils, especially ammonites and echinoderms, but others are motivated by the thrill of discovering these treasures of the Earth, or by the satisfaction that can be derived from identifying and appreciating the significance of the finds they make.

Fossils of invertebrates – animals without backbones – are not only more diverse than either vertebrate animal or plant fossils, but are also more likely to be found when collecting fossils from the majority of geological sites. Consequently, a large number of palaeontologists focus on invertebrate fossils and these fossils often dominate the holdings, if not the public exhibits, of museums. Our aim in this book is to introduce examples of the more common fossil invertebrates from around the world, as well as some rarer but scientifically significant fossils. We have set out to highlight the appreciation of fossils as the remains of once living animals, not merely as oddly shaped stones. Breathing life back into fossils is best achieved by reference to related modern species whose lifestyles and behaviours can be observed directly. Conversely, as we will attempt to demonstrate, fossils provide many insights into the origin and evolution of modern animals. As this book was not written as a textbook we have tried to avoid introducing terminology except where necessary, for example, when discussing the functions of particular features and when terminology serves as a useful shorthand for explaining differences between fossils.

Brief illustrative descriptions are included for some characteristic fossil genera at the end of each major section of text. These are not intended to be comprehensive; specialist texts should be consulted for identification purposes.

Paul D. Taylor and David N. Lewis
June 2004

# About the authors

Paul D. Taylor has worked in the Department of Palaeontology at the Natural History Museum, London for 25 years, heading the Invertebrates and Plants Division between 1990 and 2003. Author of more than 150 scientific articles, he holds a BSc in Geology and a PhD from the University of Durham. His research focuses on fossil and living bryozoans, with subsidiary interests in evolution, palaeoecology and fossil folklore.

David N. Lewis has been on the staff of the Department of Palaeontology at the Natural History Museum, London for almost 41 years and has served as a Collections Manager for invertebrates since 1990. He holds BSc and MPhil degrees from Birkbeck College, University of London. In addition to palaeontological curation, he has a research interest in fossil echinoids and has also published on trace fossils.

# Acknowledgements

Phil Hurst expertly photographed the majority of the fossils we have illustrated. Without his contribution the book would be much diminished. The following staff and associates of the Department of Palaeontology in the Natural History Museum gave generously of their time in reading sections of the text or helping to locate specimens for photography: Richard Fortey, Claire Mellish, Andrew Smith, Andrew Ross, Sarah Long, Steve Baker, Caroline Hensley, Noel Morris and Adrian Rushton. Alex Ayling and Rita Krans assisted with figure magnifications. Liz Harper (University of Cambridge) kindly read and commented on the manuscript in its entirety.

# I

# Introduction

## WHAT ARE FOSSILS?

Fossils are the remains of ancient animals and plants preserved by natural processes of burial. They come in many different shapes and sizes. Palaeontologists tend to divide fossils into four main categories: microfossils, plants, vertebrates and invertebrates. Microfossils are too small to be visible clearly with the naked eye and must be studied using a microscope. Without doubt they are the most abundant of all fossils – tens of thousands can occur in a rock of modest size – but generally pass unseen by non-specialists because of their tiny size. Fossil plants, including wood, leaves, seeds and occasionally even flowers, are locally very abundant, to the extent that they sometimes form rocks, notably coal. Vertebrates – animals with backbones – are represented in the fossil record by bones and teeth, and very occasionally by mummified or mineralised flesh. Considerable public and scientific attention has been directed towards two groups of fossil vertebrates: dinosaurs, real life monsters in scale at least, and hominids, the closest relatives of human beings. The fourth category of fossils, invertebrates, forming the

subject of this book, are the fossils most likely to be noticed and collected at geological sites by amateurs and professionals alike.

As previously defined, fossils are the remains of ancient organisms preserved through burial. An early use of the term fossil, derived from the Latin word *fossilis*, was for any object dug out of the ground. Pieces of pottery and crystals of inorganic minerals that are nowadays not regarded as fossils were once classified as such. Only organic remains fall within the modern definition of a fossil.

'Pseudofossils' are objects resembling fossils but are not true fossils. Oddly-shaped rocks formed through inorganic processes may look like fossil bones or even entire corpses of animals. For example, septarian concretions can bear a striking resemblance to tortoise or turtle shells. These structures form beneath the sea bed through an inorganic process of chemical precipitation. Shrinkage of the concretion occurs as water is forced out, and the cracks that result are filled with minerals such as calcite precipitated from percolating solutions. Their inorganic origin places septarian concretions outside the modern definition of a fossil. Nonetheless, septarian and other concretions do often contain true fossils because inorganic precipitation frequently began around organic remains preserved in

**Left** Concretion split open to reveal the fossil of a Pliocene bivalve mollusc preserved as an internal (upper fig.) and an external (lower fig.) mould. The shell itself, originally 3.5 cm wide, has dissolved away.

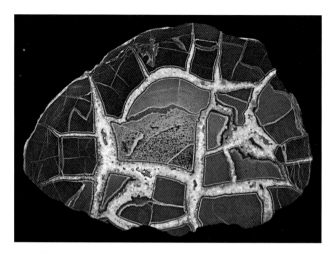

**Above** Cut and polished septarian concretion, 15 cm wide, showing the mineral-filled cracks. Although sometimes mistaken for fossils, septarian concretions have an inorganic origin.

Although by no means as widespread as the media like to portray them, fake fossils made by carving rocks into organic shapes, or by doctoring true fossils, do exist. Historically, the most famous fossil fakes are 'Beringer's lying-stones'. Johann Beringer (1667–1740) was a German academic and physician who published a book describing exquisite fossils of frogs, birds with newly lain eggs, spiders in their webs and even miniature moons and comets, all allegedly collected from a hill near Würzburg. It later emerged that these were carved stones commissioned by two malicious colleagues of Beringer's. Fake fossils are still manufactured today for commercial sale but are seldom sufficiently convincing to deceive palaeontologists and distort the progress of scientific knowledge.

the centre of the concretion. While the fact that septarian concretions are pseudofossils is well established, it is worth noting that there are still geological specimens whose identities as fossils have yet to be resolved. These enigmatic specimens are found mainly in very ancient rocks or rocks affected by high temperatures and pressures.

## WHAT KINDS OF ROCKS CONTAIN FOSSILS?

Rocks are aggregates of minerals, that is chemical compounds with distinct compositions, properties and crystalline structures. There are three main kinds of rocks: igneous, sedimentary and metamorphic. The great

## FOSSIL FOLKLORE

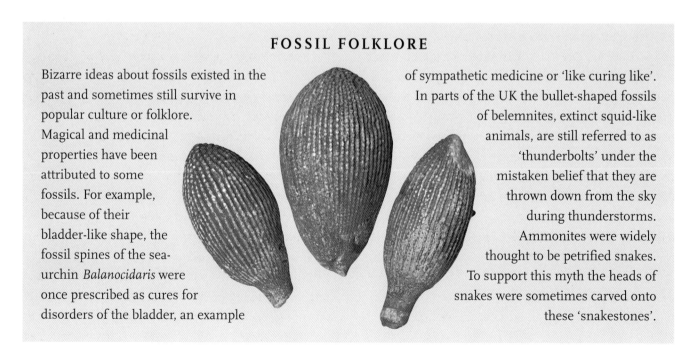

Bizarre ideas about fossils existed in the past and sometimes still survive in popular culture or folklore. Magical and medicinal properties have been attributed to some fossils. For example, because of their bladder-like shape, the fossil spines of the sea-urchin *Balanocidaris* were once prescribed as cures for disorders of the bladder, an example

of sympathetic medicine or 'like curing like'. In parts of the UK the bullet-shaped fossils of belemnites, extinct squid-like animals, are still referred to as 'thunderbolts' under the mistaken belief that they are thrown down from the sky during thunderstorms. Ammonites were widely thought to be petrified snakes. To support this myth the heads of snakes were sometimes carved onto these 'snakestones'.

majority of fossils come from sedimentary rocks that are formed by the accumulation of grains of sediment, ranging from microscopic mud particles, through silt, sand, pebbles and cobbles to large boulders. The products of weathering and erosion, these sedimentary grains are transported by wind, water currents or ice before finally being deposited. Deposition (or sedimentation) can occur in many different environments, including river channels, lake beds and deserts. However, the sea is the most important site of sediment deposition. More sediment accumulates on the sea bed than anywhere else, and it is less likely to be removed by the forces of erosion that prevail on the land surface. Therefore, a high proportion of sedimentary rocks containing fossils were deposited in marine environments.

What happens to sediments after they have been deposited? Those that are not subsequently eroded become progressively buried beneath more sediment. Burial brings about complex changes, collectively termed 'lithification' or 'diagenesis', which turn the sediment into a sedimentary rock. Compression reduces the spaces between the sediment grains and, if the sediment was originally deposited in a watery environment, drives out the water, causing a reduction in volume. The spaces between sediment grains may stay open but, over time, are more likely to become filled with mineral cements precipitated from solutions by groundwater passing through the sediment, in much the same way as stalactites and stalagmites form in caves by the slow precipitation of minerals. As the grains become tightly bound together by this cement, the sediment hardens into a rock. Lithification turns mud into mudstone or shale (if fissile), silt into siltstone, sand into sandstone, and pebbles into conglomerate or breccia (if angular). While lithification is proceeding, the fossils themselves are changing. They may become compacted and crushed, dissolved or replaced entirely or in part by precipitated minerals.

Sedimentary rocks develop various structures during their formation. Vertical or near vertical cracks called

**Above** Fossil-bearing sedimentary rocks of Jurassic age exposed in the cliffs of Normandy, France. The rocks are strongly layered or stratified.

joints form as the sediment shrinks in volume. Initially horizontal planes in the rock termed bedding planes usually represent slight pauses in deposition or changes in the type of sediment that was accumulating, for example a switch from mud to sand deposition. The presence of bedding planes gives sedimentary rocks their characteristic bedded or stratified appearance. Sedimentary rocks tend to weather and fracture more easily along bedding planes than in other orientations, revealing the fossils of animals that lived during the breaks in deposition plastered across the bedding plane. When collecting fossils it is generally advisable to split the rock parallel to bedding planes. Sediments deposited under the influence of currents, either on the land surface or under the water, often form dune structures that migrate in the direction of current flow. This produces oblique or cross bedding in the sedimentary rock, reflecting the dipping upcurrent surfaces of the dunes where the sediment is deposited. Ripple marks and mudcracks are among a host of other structures found in sedimentary rocks.

Limestones are a special kind of sedimentary rock that are often very rich in fossils. Indeed, some limestones consist almost entirely of intact or fragmented fossils. Unlike the sedimentary rocks already mentioned that comprise silicate minerals (e.g. quartz, feldspar, clay minerals), the fossils and other grains in limestones are made of carbonate minerals. The most common carbonate mineral is calcite. This particular crystalline form of calcium carbonate is utilised by many animals and plants to build their shells or skeletons. One very pure variety of limestone called chalk is extremely fine-grained and consists overwhelmingly of coccoliths, the miniscule skeletal plates secreted by a particular kind of planktonic algae. Crinoidal limestones are made of the plates of sea-lilies, a group of echinoderms, and reef limestones are made of corals and other reef-forming and reef-dwelling organisms. Oolitic limestone is a distinctive type of limestone with a texture similar to fish roe (eggs). It consists of grains called ooids, each less than 2 millimetres in diameter and with a concentric internal structure. These ooids rolled across the sea bed like tiny snowballs picking up carbonate and growing in size.

Whereas limestones provide the richest source of fossils, the best preserved fossils are generally found in clays, mudstones and shales. There are several reasons why this is the case. These 'argillaceous' rocks are very fine-grained, tend to have been deposited in tranquil conditions and are often poorly cemented, allowing sediment particles around the fossils to be easily washed away. Often argillaceous sediments formed in conditions of low oxygen, retarding the decomposition of organic remains and making fossilisation more likely. In contrast, limestones can be very hard, the fossils being firmly cemented together and difficult to extract. The coarser grained sandstones and conglomerates are porous, promoting passage of water through the rock and with it the dissolution of the fossils present.

Igneous rocks, formed by the solidification of molten material (magma), including lava extruded onto the Earth's surface, seldom contain fossils. This is not surprising in view of the high temperatures in which they formed and also their frequent solidification deep beneath the surface of the Earth where no life exists.

Animals and plants caught up in lava flows extruded onto the Earth's surface are occasionally fossilised, although in many instances they are represented only by voids in the rock left after combustion of the organic remains. Ashes ejected from volcanoes can bury and consequently fossilise animals and plants unfortunate to be caught up in ash falls. Spectacular examples of fossilisation in volcanic ash are the human remains from Pompeii in Italy that were buried during the devastating eruption of Vesuvius in AD 79.

Metamorphic rocks are formed through the action of high temperatures and pressures on existing rocks. Limestone is converted to the metamorphic rock marble, sandstone to metaquartzite, and shale to slate, schist or gneiss depending on the severity of metamorphism. Fossils in metamorphic rocks are commonly distorted as a result of shearing of the rock. With increasing temperatures and pressures the fossils in sedimentary rocks deteriorate until they become totally unrecognisable.

## HOW ARE FOSSILS FORMED?

The great majority of animal fossils comprise hard parts sufficiently resilient not to decay and disintegrate totally after the animal has died. These hard parts consist mainly of the bones and teeth of vertebrates, and the shells and skeletons of invertebrates such as molluscs and echinoids. Fossilisation happens because these skeletal components are composed predominantly of tough inorganic minerals that were employed by the living animals for protection, grinding food, or for supporting the soft parts of the body. The calcium phosphate minerals forming teeth and bones and the calcium carbonate minerals forming shells are not easily broken down by the microbes that destroy soft tissues. Furthermore, these minerals are capable of surviving the geological agents of destruction that act over the long-term after burial.

The process of fossilisation can be illustrated by the fate of a snail living on the sea bed. Following death, the soft parts of the snail's body (e.g. muscles, nervous tissues, blood vessels and various organs) immediately begin to decay, being ingested by microbes and the snail's own digestive juices. Larger scavenging animals may also contribute to this destruction, tearing off chunks of flesh on which to feed. Eventually, only the hard shell remains and this too may begin to disintegrate through dissolution in the water or as the proteins binding the inorganic minerals of the shell decompose, allowing the tiny crystals to fall apart. Reprieve from total destruction may come if the shell is suddenly buried, for example, by a storm dumping a covering of sand on the sea bed. Once buried, the shell becomes an incipient fossil. Even then, however, there is a good chance that the shell will be returned to the surface where it will once again be vulnerable to decay and destruction. For example, erosion of the sediment by currents, or the activities of animals burrowing through the sediment may result in exhumation of the shell. Some shells do remain entombed and become buried ever more deeply as more and more sediment accumulates on the sea bed.

Many animals have multipart skeletons, as in the tests of sea urchins which are made up of hundreds of individual plates. The soft tissues holding such skeletons together during life begin to decay after death, causing disaggregation of the skeleton. After disaggregation, individual plates may become dispersed, with smaller, lighter plates more likely to be carried away by currents sweeping the sea bed, a process called winnowing. Therefore, rapid burial, important in any form of fossilisation, is especially critical for preserving animals with multipart skeletons.

Exceptionally, some or all of the unmineralised parts of animals can be fossilised. Spectacular examples include insects in amber, and the skin and hair of mammoths frozen in permafrost. Preservation of soft parts may be due to a variety of circumstances. Extremely rapid burial is always necessary. Often organic decay is suppressed by extremely cold conditions, as with the mammoths, or chemical environments unfavourable to the microbes that normally degrade soft tissues, as with insects trapped in tree resins that will ultimately form

## TRACE FOSSILS

This special category of fossils is recognised for traces of the activities of animals rather than their fossilised bodies. Trace fossils include the tracks left by animals moving across the land surface, burrows excavated by worms and crustaceans in sand, and holes drilled by particular kinds of molluscs, barnacles and worms into shells or limestone rock. Animal dung, known as coprolite, represents another category of trace fossil. Whereas trace fossils can tell us a lot about the behaviour of animals, such as the speed of running in the case of the tracks left by dinosaurs, it may be difficult or impossible to ascertain the exact identity of the trace maker. On the one hand identical trace fossils can be made by completely unrelated groups of animals behaving in similar ways, while on the other hand particular animals may make two or more kinds of trace fossils depending on the activity in which they were engaged. Therefore, a separate system of naming trace fossils has been developed. Trace fossils are given ichnotaxon names. For example, *Gastrochaena* is the name of a common genus of rock-boring bivalve molluscs whose shells are frequently found as body fossils. Fossilised holes made by *Gastrochaena* are given the trace fossil name *Gastrochaenolites*. However, there is no one-to-one relationship between trace and body fossil because *Gastrochaenolites* can also be made by another bivalve called *Lithophaga* as well as other animals.

**Above** A boring made by a sponge into a shell during Cretaceous times has here been infilled with flint. Subsequent dissolution of the shell has revealed galleries (each less than 2 mm in size) linked by tunnels constituting the trace fossil *Entobia*.

amber. Environments having low oxygen levels, such as stagnant muds on lake beds, often favour soft part preservation for the same reason. Also important in soft part preservation is the activity of other microbes whose activities cause the precipitation of phosphate minerals. These minerals can grow very rapidly to replace in fine detail the soft tissues of the dead animal.

Deposits containing exceptional fossils are known by the German term '*Fossil-Lagerstätten*' (singular '*Lagerstätte*'). They provide 'windows' on parts of the biosphere not normally represented in the fossil record; many fossils are known only from their occurrences in *Lagerstätten*. Among the best known *Lagerstätten* are the Burgess Shale (Cambrian, Canada), Hunsrück Shale (Devonian, Germany), Mazon Creek (Carboniferous, USA) and Solnhofen Limestone (Jurassic, Germany).

There have been recent scientific claims that DNA, the chemical blueprint of life, may survive in some fossils, especially insects in amber. The publicity given to these claims was heightened by the release of the movie *Jurassic Park* (1993) in which dinosaur DNA was supposedly extracted from amber-preserved mosquitos that had fed on dinosaur blood. In the ingenious plot of *Jurassic Park* this fossil DNA was used to recreate living dinosaurs. Unfortunately, there is little evidence that DNA really can survive for millions of years of geological time, even in a highly fragmented state, and the recreation of species that became extinct 1 million or more years ago from fossil DNA may never be possible.

## FOSSIL PRESERVATION

There are many different modes of fossil preservation depending on the nature of the original remains and subsequent changes to them after burial. For example, the shells of snails and other molluscs made of the mineral aragonite are often dissolved entirely by percolating water, leaving a void in the rock. This void is a natural mould – a negative impression – of the fossil shell. It has two parts: an external mould of the outer surface of the shell, and an internal mould (or steinkern) of the inner surface of the shell. The mould can be subsequently filled by new minerals precipitated from solution to give a natural cast of the fossil. The cast replicates the shape of the snail shell but, of course, does not reproduce the original internal fabric of the shell which will have been lost during dissolution. In some instances, however, shell material may be gradually and progressively replaced by new minerals such that traces of the original internal structure of the shell are preserved. Shells in limestones are sometimes replaced, with or without conservation of internal fabric, by silica minerals. Silicification can be highly advantageous to palaeontologists because it enables the fossils to be extracted by treating the rock with acids: the carbonate minerals forming the matrix of the rock dissolve whereas the silicified fossils are left behind as an insoluble residue that can be rinsed and picked over.

## DATING FOSSILS

The Earth is immensely old. Current estimates are that it was formed some 4500 million years ago from the condensation of a dust cloud encircling the Sun. The oldest rocks on Earth contain no fossils, but in rocks of about 3500 million years old the first traces of fossilised life have been found. For roughly the first 4000 million years of Earth history the only fossils known are of microbial organisms such as bacteria. During the interval from about 550 million years ago to the present day, ever increasing numbers of more complex and larger organisms evolved on Earth and left their remains in the fossil record.

But how old are the youngest fossils? This depends on the exact definition we use for the term fossil. In a broad sense, a fossil can be any animal or plant that has been buried naturally. Therefore, the carcass of a sheep caught up only yesterday in flood waters and buried in mud when the flood subsided can be regarded as a fossil. Palaeontologists commonly apply an arbitrary age for the youngest fossils, often only including those remains greater than 10,000 years old. Remains of organisms in the grey area between the sheep mentioned above and true fossils are sometimes referred to as 'sub-fossils'.

How do we determine the age of rocks and of the fossils they contain? This is usually accomplished through radiometric dating techniques. Chemical elements tend to have several isotopes that differ in the number of neutrons contained in the nucleus of the atom. Usually each element has one common isotope along with several rarer isotopes which may be stable or unstable. With time, unstable isotopes gradually decay, emitting radioactivity in the process. The important point about radioactive decay is that it occurs at a constant rate, which may be expressed as the half-life: the time taken for half of the original quantity of the isotope to decay. For example, an isotope of uranium called uranium[235] decays to lead[207] at a rate such that half of the original quantity of uranium[235] in a rock will have disappeared by the time the rock is 713 million years old. By measuring the amount of uranium[235] relative to the stable isotopes it is possible to calculate the age of the rock. Not all rocks are suitable subjects for radiometric dating. Unfortunately, sedimentary rocks are, in general, poorer for this purpose than igneous rocks. Lava flows interbedded with sedimentary rocks are consequently often a target for geologists wishing to know the age of the sedimentary rocks. Imagine a fossil-bearing sandstone sandwiched between two lava flows radiometrically dated at 85 and 90 million years old. The fossils in the sandstone must be somewhere between 85 and 90 million years old.

Fossils by themselves do not tell us the absolute ages of rocks, for example, whether a rock is 100 or 150 million years old. However, they are of great value in the determining relative ages of rocks, for example, whether rock A is older, younger or the same age as rock B.

Using fossils to determine relative ages of rocks is possible because the animals and plants living at different times during Earth's history have changed. Therefore, particular species indicate particular ages. There is a close parallel with methods of relative dating used in archaeology. Archaeologists excavating ancient settlement sites are able to determine the age of habitation by reference to the kinds of buried pottery and other cultural artefacts that they find. Furthermore, just as objects that were used by human cultures for short periods of time are more useful to archaeologists for dating than are objects with a longer history of use, so palaeontologists find that some fossils are better than others for estimating the ages of rocks. Ideal fossils comprise remains of animals and plants that evolved rapidly so that new species followed one another in quick succession. It is also important that they are geographically widespread species and are present in sufficient abundance to be routinely discovered by palaeontologists.

The use of fossils for ascertaining the relative ages of rocks has a long history. In the UK, it was pioneered by William Smith who utilised fossils to recognise the outcrop patterns of strata of different ages in order to construct his famous geological maps. Ever more precise and finer schemes of subdivision have been developed since William Smith's work 200 years ago. For example, the Jurassic period in the UK is now divided into more than 70 zones, each on average 1 million years long, and defined by the occurrence of a particular species of ammonite. Sedimentary rocks present in geographically distant places can be correlated by the discovery in them of the same species of zonal fossils. The application of fossils in this way is known as biostratigraphy and has been key to the exploration of natural resources such as oil.

## THE GEOLOGICAL TIMESCALE

The geological timescale is a kind of world map of time made up of major divisions that are subdivided into successively smaller units, in much the same way as a

| EON | ERA | Period | | Epoch | |
|---|---|---|---|---|---|
| Phanerozoic | Cenozoic | Quaternary | | Recent | 0 |
| | | | | | 0.01 |
| | | | | Pleistocene | |
| | | | | | 1.8 |
| | | Tertiary | Neogene | Pliocene | |
| | | | | | 5.3 |
| | | | | Miocene | |
| | | | | | 23 |
| | | | Paleogene | Oligocene | |
| | | | | | 34 |
| | | | | Eocene | |
| | | | | | 56 |
| | | | | Paleocene | |
| | | | | | 65.5 |
| | Mesozoic | Cretaceous | | | 146 |
| | | Jurassic | | | 200 |
| | | Triassic | | | 251 |
| | Palaeozoic | Permian | | | 299 |
| | | Carboniferous | | | 359 |
| | | Devonian | | | 416 |
| | | Silurian | | | 444 |
| | | Ordovician | | | 488 |
| | | Cambrian | | | 542 |
| Precambrian | Proterozoic | Ediacaran | | | 600 |
| | | | | | 2500 |
| | Archaen | | | | 4000 |
| | Hadean | | | | 4500 |

**Left** Geological timescale with ages in millions of years. Note that the vertical scale is uneven – for example, the Precambrian was of much longer duration than the Phanerozoic.

geographical world map has continents, countries, counties, cities, etc. The Precambrian, a huge chunk of geological time spanning the first 4000 million years of Earth history, is followed by the Phanerozoic in which most non-microscopic fossils are found. There are three major divisions, or eras, of the Phanerozoic: Palaeozoic, Mesozoic and Cenozoic (sometimes written Caenozoic). These in turn contain subdivisions called periods. The names given to the geological periods by the geologists who originally established them have various origins. Some are derived from geographical regions (e.g. Devonian after the county of Devon in England, Permian after the Russian city of Perm) but others have different roots (e.g. Ordovician after the Ordovices, Silurian after the Silures, both ancient Welsh tribes; Triassic because of the threefold sequence of rocks of this age found in mainland Europe). Periods themselves are subdivided into stratigraphical stages, usually named after particular places (e.g. Bathonian after Bath in southern England, Cenomanian after Le Mans in France). Details of the geological timescale are under constant revision. For example, during 2004, a new geological period, the Ediacaran, was formally introduced for the last major interval of the Precambrian.

Palaeontologists like to place their fossils into as precise a timeframe as possible. This is why it is always important to record exact details of where and at what particular level in the rock sequence particular fossils are collected. The better this information, the more useful it is in understanding the evolution of life through time.

## INVERTEBRATES

Human beings, together with all other mammals, birds, reptiles, amphibians and fish, have an internal skeleton comprising bones and cartilage. One of the key components of this skeleton is the backbone, or vertebral column, which provides the collective name 'vertebrates' for such animals. Vertebrates are a natural biological group, that is they consist of an ancestral species (which existed some 500 million years ago) plus all of its

descendants. In contrast, invertebrates, the subject of this book, are not a natural biological group; they are a so-called 'paraphyletic' group comprising what remains of the animal kingdom after the vertebrates have been subtracted. Nonetheless, invertebrates are a recognisable entity distinguished, as their name implies, by the absence of a vertebral column.

Think of any animal that is not a mammal, bird, reptile, amphibian or fish, and it will be an invertebrate. Taken together, the vertebrates comprise only one of the 30 or so major groupings of animals referred to as a phylum; all of the remainder are invertebrates. Three invertebrate phyla are particularly familiar to humans because they include terrestrial species that inhabit the land surface or fly in the sky: Arthropoda (e.g. spiders and flies), Annelida (e.g. earthworms) and Mollusca (e.g. slugs and snails). However, only in aquatic environments, especially the sea, does the full diversity of invertebrates becomes apparent. Here can be found invertebrate phyla such as Porifera (sponges), Cnidaria (e.g. corals and jellyfish), and Echinodermata (e.g. starfish and sea urchins), as well as many additional species belonging to the Arthropoda (e.g. crabs and shrimps), Annelida (e.g. lugworms) and Mollusca (e.g. mussels and octopus). A host of less familiar phyla also live in the sea, some common as fossils (e.g. Hemichordata, Brachiopoda and Bryozoa) but others rare (e.g. Priapulida and Sipunculida).

Invertebrate phyla are defined by their distinctive anatomical body plans. Each phylum has a unique organisational pattern of the tissues and organs forming the body, and a characteristic sequence of embryological development. The body plans of most, possibly all, invertebrate phyla evolved at least 550 million years ago; rocks of Cambrian age contain the first fossil evidence of many of invertebrate phyla with mineralised skeletons as well as some soft-bodied phyla. Apart from sponges and cnidarians, no extant (still living) invertebrate phyla have unequivocal representatives in the Precambrian fossil record, although the peculiar fossils found in the

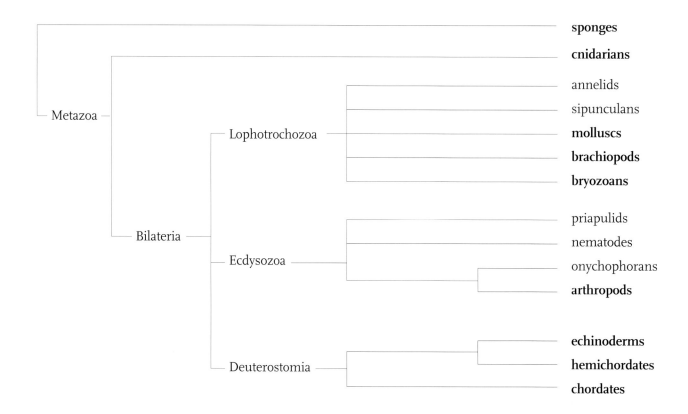

```
                                                                    sponges

                                                                    cnidarians

                                                        annelids

                                                        sipunculans

           Lophotrochozoa                               molluscs

                                                        brachiopods

                                                        bryozoans

 Metazoa

                                                        priapulids

                                                        nematodes

           Ecdysozoa                                    onychophorans

 Bilateria                                              arthropods

                                                        echinoderms

           Deuterostomia                                hemichordates

                                                        chordates
```

**Above** Evolutionary relationships between the main phyla of invertebrate animals mentioned in this book. Those groups of particular importance as fossils are in bold.

latest Precambrian Ediacaran period are considered by some palaeontologists to include primitive representatives of modern phyla. The abrupt first appearance of several invertebrate phyla in the Cambrian is referred to as the 'Cambrian Explosion'. Some scientists believe that the Cambrian Explosion corresponds with the origination of these phyla, whereas others interpret it as merely representing the time of evolution of fossilisable hard skeletons, with the phyla themselves having originated further back in Precambrian times. Evidence favouring this second alternative has come from molecular sequencing of present-day species belonging to the various invertebrate phyla. The differences evident in these molecular

sequences imply that divergence between the phyla occurred as long as 1000 million years ago. However, this calculation rests on the validity of something called the 'molecular clock hypothesis' which makes the contentious presumption of a constant rate of molecular evolution through time. It also depends on the assumption that molecular divergence corresponded exactly with the acquisition of the characteristic body plans we now use to distinguish the phyla.

Understanding the inter-relationships between the invertebrate phyla has exercised the minds of zoologists and palaeontologists for more than 150 years, without the emergence of a clear concensus. New data from gene sequences suggests that the so-called bilaterians, a group of animals excluding the sponges and cnidarians, divide into three large groups called lophotrochozoans, ecdysozoans and deuterostomes. However, it is far less clear how the phyla within these groups are inter-related.

# INVERTEBRATE DIVERSITY THROUGH TIME

Documenting and interpreting patterns of changes in the diversity of invertebrate life through geological time is a major focus of interest for palaeontologists. As it is not usually practical to compile data for species, most studies investigate such evolutionary patterns at the level of families or genera. A widely-reproduced diagram summarising the diversity pattern for marine invertebrate families originated from the work of the late Jack Sepkoski, a prominent palaeontologist who was based at the University of Chicago. The 'Sepkoski curve' shows that the number of families increased rapidly during the Cambrian Explosion, with diversity continuing to rise into the succeeding Ordovician period. Diversity more or less levelled out for the remainder of the Palaeozoic, except for drops in the late Ordovician and late Devonian. These drops represent the first two of five geologically sudden episodes of mass extinction that decimated life in the sea (and on land). The third mass extinction, at the end of the Permian, was the largest of them all. More than 95 per cent of marine species may have become extinct and there is also a huge drop in family diversity. The fact that this gigantic mass extinction marks the boundary between the Palaeozoic and Mesozoic eras is not a coincidence; the striking change in the fossils found at this level was the reason for making it the boundary between the two eras.

Diversity rose again after the Permian mass extinction only to be knocked back by another mass extinction at the end of the Triassic period. The fifth mass extinction at the end of the Cretaceous, the so-called KT mass extinction, has achieved fame as it not only accounted for the last species of dinosaurs, but also because of mounting geological evidence for the collision of a large asteroid with the Earth that may have caused the extinction. In the long term, neither the end-Triassic nor the KT mass extinctions had much effect on the inexorable rise in the number of marine invertebrate families through the Mesozoic and Cenozoic. The fossil record towards the end of the Cenozoic contains roughly twice as many families as were present in the Palaeozoic.

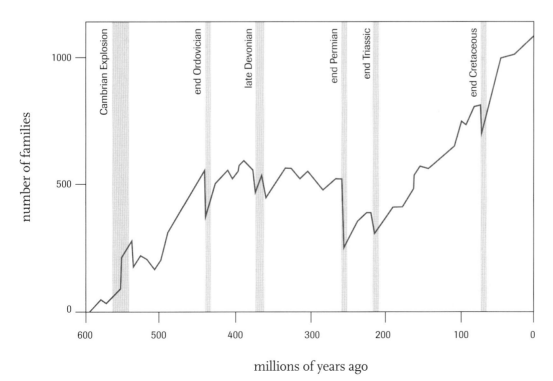

**Left** Changes in the diversity of invertebrates living in the sea are evident from the Sepkoski curve showing the number of families recorded in the fossil record through geological time. The Cambrian Explosion and five major mass extinctions are indicated.

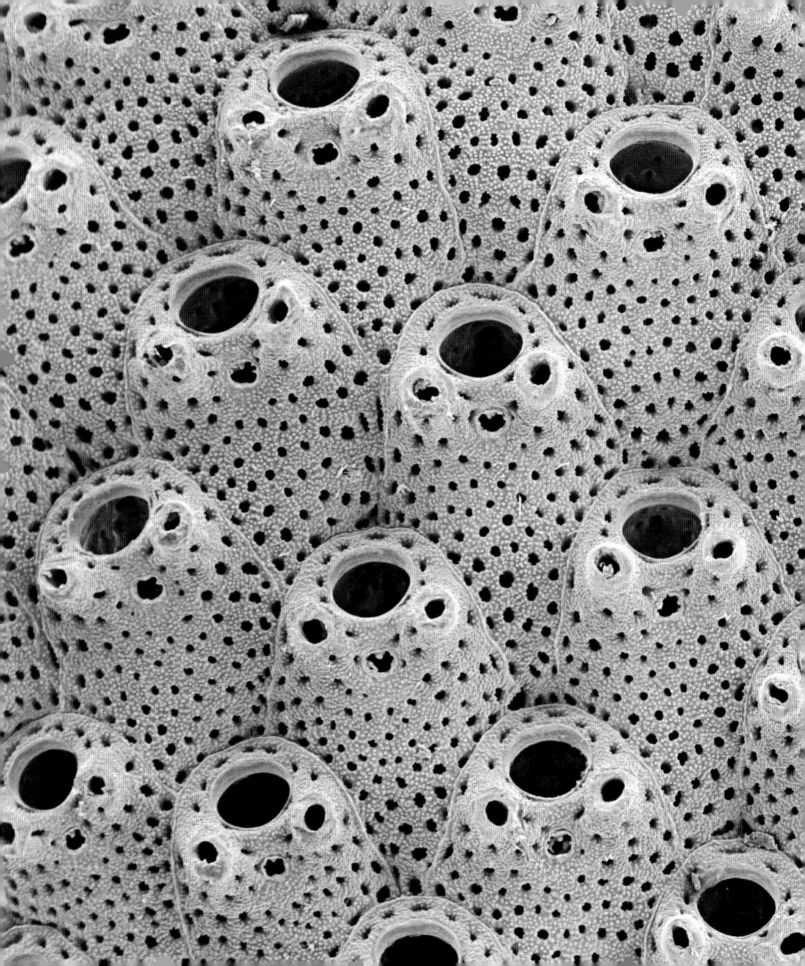

# 2

# Living in colonies

IN HUMANS, identical twins are uncommon, conjoined twins ('Siamese twins') even rarer. Identical twins are clones, that is they have exactly the same genetic make-up. Humans, and indeed all other animals living on land, are unitary animals. However, in watery environments, particularly the sea, a large number of animal species, including reef corals, are colonial; they exist as multiples of conjoined, clonal units. Each individual unit of the colony is known as a zooid. Colonial animals differ fundamentally from unitary animals, including humans, in several ways. In particular, animal colonies are often shaped more like plants, resembling small trees, with others being similar to the lichens that grow on rocks, walls and gravestones. It is not surprising that early naturalists were perplexed by colonial animals, not knowing whether they should be classified as animals or plants. This confusion is evident in the name formerly applied to them – zoophyte – literally meaning animal (zoo-) plant (-phyte).

## COLONIAL ANIMALS

Several phyla that are common in the fossil record contain colonial species. Many cnidarians are colonial, as are all species of bryozoans and the graptolite hemichordates, while sponges have many colonial attributes even though their primitive tissue organisation means that distinct zooids are seldom recognisable. The planktonic graptolites aside, these colonial animals are usually benthic and sessile, that is they live on the sea bed and are fixed permanently in one position throughout their adult life. For example, the corals which inhabit and indeed construct the Great Barrier Reef of Queensland are animal colonies that cement themselves to the reef surface. Each coral colony consists of many clonal polyps – the fleshy parts of the zooids – and colony growth occurs by the addition of new zooids.

## FEEDING

Marine animals fixed in one position feed in a different way from the animals that cohabit the land surface with humans. Because they cannot move to search for, gather and capture food, they must rely on planktonic food being brought into their vicinity by environmental water currents, and then use tentacles or some other filtration mechanism to capture the plankton. Feeding by this method is referred to as suspension feeding. There are two types of suspension feeding, passive and active, depending on whether the animal relies entirely on

**Left** Modularity in the colonial skeleton of the modern bryozoan *Inversiula*. The modules (zooids) are less than a millimetre long and have an oval opening through which the tentacles emerged during life.

environmental currents or supplements these by creating water currents of its own. Perhaps the closest that land-dwelling animals come to suspension feeding can be found in spiders that weave webs to capture passing flies. From a human perspective, suspension feeding is a bizarre way of obtaining food; it is almost like sitting permanently on a stool in one of those sushi bars where dishes of food pass by on a conveyor belt and can be grabbed and eaten. Suspension feeding is not the sole preserve of colonial animals; many unitary animals, such as brachiopods and oysters, are also suspension feeders. However, suspension feeding is just about the only type of feeding employed by colonial animals, although reef corals also obtain some of their nutrition from the symbiotic algae present in their tissues.

GROWTH AND LIFESPAN

Colonial animals grow principally by adding new zooids to the colony, a process referred to as 'budding' and which usually occurs at the outer perimeter of the colony. For example, zooidal budding at the tips of the branches of tree-like colonies causes the branches to lengthen, much as the branches of a real tree grow upwards from the terminal shoots. Individual colonies can contain tens, hundreds, occasionally thousands and sometimes even millions of zooids. In a few species colonies grow until they reach a constant size and no longer, but in most species this is not the case and instead colonies continue to grow by budding new zooids for as long as they live. Size varies greatly between colonies of the same species in the latter case. Not only size but also shape may vary considerably within each species. Some colonial species are very 'plastic' in shape. This plasticity is often related to environmental conditions during colony growth, for example, branches developing in regimes of stronger current flow where plankton is more plentiful may grow faster, making the colony lop-sided.

Another peculiarity of colonial animals is that colonies can go on living after some or most of their constituent zooids have died. This phenomenon is called 'partial mortality'. Because the lifespan of individual zooids is typically much shorter than that of the colony as a whole, the older zooids in living colonies are frequently dead. The zone of dead zooids close to the colony origin, for example in the basal branches of tree-like colonies, is referred to as a 'necromass'. Although dead, the necromass zooids in tree-like colonies are functionally important in that their skeletons support the younger, actively feeding zooids near the branch tips. When examining fossil colonies it is easy to be misled into thinking that all of the zooids were feeding immediately prior to the death of the colony, whereas often only a minority of zooids were in fact active. Zooids can be lost entirely from living colonies, for example if a branch of a tree-like colony snaps off as a result of a storm surge. Along with the often very variable growth rates of colonial species, such losses mean that the size of a colony may be a poor indicator of its age, providing another contrast with unitary animals where size is usually a reliable indicator of age.

Individual zooids in colonies may also die as a result of attacks by predators. The predators of colonial animals in some ways act more like parasites in that they do not always kill the entire colony but instead take single zooids or small groups of zooids at a time. Indeed, while predators are devouring zooids in one part of the colony, new zooids may be being formed elsewhere at a rate equal to or even exceeding those lost through predation.

POLYMORPHISM AND PHYSIOLOGY

Colonies exhibit varying degrees of interdependency between their constituent zooids, although quite how much is difficult to gauge. Many animal colonies contain zooids that lack the ability to feed. These non-feeding zooids must be nourished by their feeding neighbours. But why should colonies contain non-feeding zooids? This is because zooids in some colonial species have become specialised for different functions, such as reproduction, and defending or cleaning the colony. Such 'polymorphic' zooids have distinctive shapes reflecting their differing functional roles. For example, zooids used to brood larvae prior to their release may be bulbous so they can accommodate as many developing larvae as possible. Polymorphic zooids can usually be recognised in fossils because their functional specialisation is reflected by modification of the skeleton. The function of polymorphs in extinct colonies can often be inferred by comparison with living relatives in which the behaviour of the polymorphs can be observed. However, the polymorphic zooids of some extinct groups have no close analogues among living species and their function remains unclear.

Polymorphism is rare in corals but well-developed in most other colonial animals, reaching its acme in a group of cnidarians called siphonophores. These soft-bodied animals, scarcely represented in the fossil record, include the notorious Portuguese Man-o-War. Venomous colonies of this jellyfish-like creature float around the world's oceans, dangling their tentacles armed with stinging cells able to paralyse any animals unfortunate enough to touch them. The gas-filled float, or pneumatophore, is a very highly modified zooid, while the tentacles comprise several types of polymorphs, including ones used in reproduction (gonozooids), food capture (autozooids) and digestion (gastrozooids).

Providing nourishment to non-feeding zooids, as well as transferring energy to the parts of the colony where new zooids are being budded or larvae are brooding, demands that the zooids in colonies are linked by soft tissues and not just joined by their skeletons. Pores in the skeletal walls between the zooids of fossil colonies may testify to the former existence of soft tissue connections, and zooids can also be linked by an external covering of soft tissues on the outside of the colony above the skeleton. Relatively little is known about the physiology of animal colonies but it is apparent that the extent to which zooids are 'integrated' varies according to species. In poorly-integrated colonies individual zooids seem to be almost independent, whereas in well-integrated colonies they are subservient to the colony as

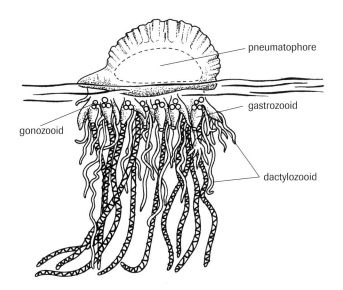

**Above** The Portuguese Man-o-War *Physalia* exhibits extreme polymorphism, with zooids of radically different shape performing separate functions.

a whole. Nerves running between zooids make co-ordinated responses possible.

A feature of colonial animals is the small size of the zooids compared to individuals in related unitary species. Colonial corals, for example, have zooids which, on average, are substantially smaller than non-colonial corals. Small zooids have a large surface area compared to volume (area increases as a squared function of linear dimension, whereas volume increases as a cubed function). This confers physiological advantages, notably in respiration and excretion. All of the cells may be close enough to the surface of the zooid that simple diffusion is adequate to supply respiratory oxygen and remove carbon dioxide and other waste products. Circulatory and excretory systems may therefore be unnecessary. On the other hand, being large confers its own advantages; large animals are less likely to be killed by predators or environmental accidents. Colonial animals get the best of both worlds, reaping the physiological benefits of small zooid size while potentially having a large total size as a result of the large number of constituent zooids. They are the 'Little Big Men' of the animal world.

## COLONY FORMATION

New colonies are generally formed when free-swimming larvae settle, metamorphose and begin budding zooids. Most such larvae are products of sexual reproduction. Therefore, the life cycles of animal colonies include two principal components: sexual reproduction for dispersal and the foundation of new colonies, and asexual reproduction for colony growth. An alternative way of making new colonies adopted by some species is through the fragmentation of existing colonies. This is especially important in corals with delicate branches prone to breakage, especially during storms. Fragments of the colony may be swept away, many perishing but some surviving, resuming growth and founding a new colony which has exactly the same genetic make-up as the parent colony. Fragmentation is a type of clonal propagation, akin to a gardener cultivating new plants from cuttings. Regenerative growth from fractured edges of skeletons is sometimes recognisable in fossil colonies, testifying to the occurrence of colony formation through fragmentation in the geological past.

A further peculiarity of colonial animals is that they are able throughout their lives to produce new reproductive (germ) cells directly from the normal (somatic) cells of the body. These germ cells mature into the eggs and sperm used in sexual reproduction. In contrast, the germ cells of non-colonial species are typically produced only during the earliest stages of the development of the individual, after which time the two cell types (somatic and germ) follow independent lines. Because of this separation, mutations occurring in somatic cells are not passed on to the germ cells in non-colonial animals. Somatic mutations can, however, be transmitted to the germ cells in colonial animals. Very little is known about the implications of this difference between colonial and non-colonial species. It has been speculated, however, that if two parts of a colony develop different genetic compositions through mutation of the somatic cells, they may effectively compete with one another, the winner passing on more of its 'good' genes to the new germ cells and thence to the offspring.

# CNIDARIA

The Cnidaria, pronounced with a silent 'c', is an important phylum that includes jellyfish, sea anemones and corals. A key feature of cnidarians is the presence of specialised stinging cells called nematocysts. Employed in feeding and defence, nematocysts act like miniature harpoons, shooting out a dart tethered by a thread that can harpoon prey or enemies. These rather sophisticated structures belie the simple organisation of the cnidarian body. Cnidarians lack the coeloms (fluid-filled body cavities) found in most invertebrates, are without blood vessels or major organs such as kidneys, and have guts with only one opening that serves as both mouth and anus. Their tissue organisation is diploblastic; the walls of the body possess two layers of tightly-packed cells, an outer ectoderm and an inner endoderm. Between these is a mesoglea comprising stringy cells and a nerve net. Viewed from above, cnidarians usually exhibit radial (or near radial) symmetry in their body plans. This is evident, for example, in the continuous circle of tentacles surrounding the central mouth of a sea anemone. It is also seen in the spoke-like partitions (mesenteries) that subdivide the central body cavity (enteron) where digestion of food takes place. In corals these mesenteries may be calcified to form the septa that are conspicuous features in fossil examples.

There are thought to be more 9000 species of cnidarians living today. Most inhabit the sea but some are found in freshwater, such as *Hydra* which features in many invertebrate zoology texts as a 'model' example of a cnidarian (or coelenterate as cnidarians are known in older publications). Three large classes of cnidarians are recognised. The Hydrozoa include the 'fire coral' *Millepora* and some peculiar jellyfish-like colonial animals called siphonophores. Also belonging to the Hydrozoa are the sea-firs (hydroids) that are commonly washed ashore around the coast, entangled among seaweeds. They can be found for sale in seaside souvenir shops as 'everlasting plants', artificially dyed green in colour. True jellyfish belong to the second class, the Scyphozoa. The third class, Anthozoa, comprises sea anemones, sea pens and corals

Fossil examples of all three cnidarian classes are known but the great majority of fossil cnidarians are anthozoan corals which secrete resistant calcium carbonate skeletons. Corals are remarkable animals. They are responsible for building the Great Barrier Reef, reputed to be the only living structure visible from space. In this and other reefs, corals provide the habitat for the greatest diversity of life found anywhere in the seas of our planet. One estimate suggests that the world's coral reefs may harbour as many as 1 million species. They are very much the marine equivalent of tropical rain forests

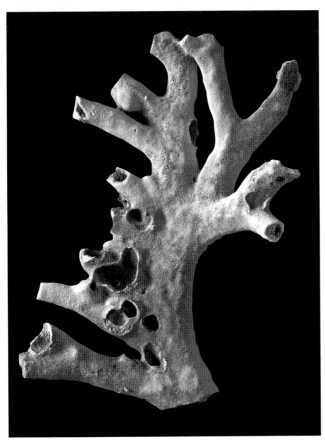

**Above** Branching colony, 11 cm in height, of the hydrozoan *Millepora* from the Pleistocene of Aldabra in the Indian Ocean.

in terms of habitat provision for other species. Coral reefs teem with vibrant and colourful life, making them targets for ecotourism as well as for conservation. The ability of reef corals to grow robust skeletons at a very rapid rate depends on the symbiotic relationship they have with algae called zooxanthellae. These algae live protected within the soft tissues of the coral, in turn providing the coral with organic compounds (glycerol, glucose and the amino acid alanine) for nutrition. While most reef corals retain the ability to feed on small animals using their stinging cells, some species are more dependent nutritionally on their symbiotic zooxanthellae. Furthermore, the photosynthetic activity of the zooxanthellae releases carbon dioxide which assists the secretion of the calcium carbonate skeleton made by the corals, allowing a faster growth rate. Reef corals live only in warm seas, where the temperature of the water does not fall below 18°C, usually at equatorial latitudes between 30°N and 30°S. They require clear water less than about 90 m (300 feet) deep. Below this depth insufficient light penetrates to satisfy the photosynthetic needs of the zooxanthellae. Another characteristic of coral reef environments is the low level of nutrients in the water. When placed under environmental stress, some corals expel their zooxanthellae and 'go it alone'. High temperatures can also bring about death of the zooxanthellae, leading to 'coral bleaching', an effect of global warming that is causing great concern.

Not all corals contain symbiotic zooxanthellae. This is true in tropical reefs as well as in corals that live in colder waters such as the loosely-branched colonial coral *Lophelia* that forms reef-like structures in deep waters of the North Atlantic. How to recognise which fossil corals had zooxanthellae and which did not is a contentious issue for palaeontologists. Although the morphology of the skeleton can offer helpful clues, a different line of evidence is provided by the isotopic composition of well-preserved fossil skeletons. Because zooxanthellae prefer to use for their metabolism the more common, lighter isotope of carbon ($^{12}$C) rather than the less common heavier isotope ($^{13}$C), the excess $^{13}$C becomes incorporated into the coral skeleton. Slightly higher proportions of this heavy isotope in the skeleton contribute to a so-called 'vital effect' which can be detected using precise analytical equipment. Fossil corals exhibiting a vital effect are thought to have had symbiotic zooxanthellae, those showing no vital effect are inferred to have lacked such symbionts. A vital effect implying the former presence of zooxanthellae has been found in skeletons of scleractinian corals dating back to the Triassic, about 200 million years ago.

There is no concensus as to whether Palaeozoic corals possessed zooxanthellae. One line of evidence suggesting that they lacked zooxanthellae is growth rate. Sclerochronology is the study of features in the skeleton that record the passage of time, very much as the growth rings of a tree do in dendrochronology. In modern scleractinian corals X-radiography can be used to detect alternating bands of high and low skeletal density, each couplet indicating one year of skeletal growth. Modern reef corals with zooxanthellae may grow 10 cm (4 inches) or more per year. For most fossil corals X-radiography cannot be used because there is insufficient contrast between the coral skeleton and the calcite cement typically filling the spaces within the skeleton, so that differences in skeletal density do not stand out. Instead, polished sections are cut to reveal the density banding and allow estimates to be made of growth rate. In combination with some other methods, this has shown that Palaeozoic corals seldom grew by more than 10 mm (0.4 inches) per year, suggesting that the zooxanthellae promoting rapid growth may have been lacking. Furthermore, the growth rate of Palaeozoic corals does not decrease in specimens collected from rocks inferred to have been deposited at greater depths. This is contrary to expectations for corals harbouring zooxanthellae that would have been inactive in lower light levels, thereby slowing down the growth of the coral.

**Above** *Cyclomedusa* from the latest Precambrian (Ediacaran) of South Australia. The slab of sandstone bearing this apparent jellyfish has a visible width of about 8 cm.

Despite the dominance of corals in the cnidarian fossil record, the most ancient cnidarians are soft-bodied species from the latest Precambrian, about 600 million years old. These ancient cnidarians are part of the famous Ediacaran biota, first discovered in the Ediacara Hills of South Australia, but since found in many other parts of the world in rocks of approximately the same age. Preserved as impressions in fine-grained sandstones, the interpretation of Ediacaran fossils has been controversial. An unorthodox view is that these fossils represent a group of organisms that became totally extinct before the Cambrian explosion of groups with hard skeletons. According to this interpretation they were a kind of 'false start' in the evolution of large-sized organisms. Alternatively, the Ediacaran fossils may be primitive, soft-bodied representatives of phyla that diversified during the Cambrian and are still living today, including cnidarians. Certainly, some Ediacaran fossils strongly resemble jellyfish, while others look like sea pens. If these attributions are correct then at least two cnidarian classes (Scyphozoa and Anthozoa) had evolved by late Precambrian times.

Recent finds of fossils interpreted as corals in Cambrian rocks have extended the range of these mineralised anthozoans back by one geological period from the Ordovician which was previously thought to mark their first appearance. Nonetheless, it was not until the Ordovician that corals became abundant members of sea bed communities. Most Palaeozoic corals belong to two groups – Rugosa and Tabulata – both first appearing in the Ordovician. Rugose corals may be either solitary or colonial; about two-thirds of rugose genera are solitary, the rest colonial. However, all tabulate corals are colonial. Neither rugose nor tabulate corals survived the great Permian mass extinction that decimated so much life on Earth. Younger corals, including the reef-building corals found in modern seas, belong mostly to a third group, the Scleractinia. These are thought by most scientists to have evolved from soft-bodied anemone-like anthozoans that independently acquired a hard skeleton, rather than directly from one of the Palaeozoic coral groups. A major difference between Palaeozoic rugose and tabulate corals and post-Palaeozoic scleractinians is evident in the mineralogy of their skeletons. The mineral calcite forms the skeletons of Palaeozoic corals, whereas aragonite occurs in post-Palaeozoic corals. This is one of the main reasons for supposing that Palaeozoic and post-Palaeozoic corals are not directly related. A consequence of the mineralogical difference is that Palaeozoic coral fossils, despite their greater age, tend to be better preserved than Mesozoic and Cenozoic coral fossils. This is because calcite is a stable form of calcium carbonate, whereas aragonite nearly always dissolves in pore waters that permeate even the most solid rocks. The aragonite skeletons of Mesozoic and Cenozoic corals are often entirely dissolved, leaving hollow moulds. Sometimes a geode-like filling of coarse sparry calcite, with or without

**Above** *Fungia* is a free-living scleractinian coral. In this large example, almost 10 cm in diameter, from the Pleistocene of Yemen, the concave underside (upper fig.) is fouled by serpulid worms and bryozoans.

traces of the original coral structure, results from the combined action of dissolution and precipitation from pore waters. Paradoxically, ancient reefs of Palaeozoic age, formed by calcitic corals and inhabited by calcitic brachiopods, bryozoans and sponges, can be in more pristine condition than younger reefs of Mesozoic and Cenozoic age with corals and associated gastropods and bivalves composed of dissolution-prone aragonite.

Solitary corals vary in shape but many are are either conical or disc-shaped. Conical corals, particularly Palaeozoic rugosans, are often curved so that they resemble a horn. The pointed end of these 'horn corals' represents the oldest part where growth originated, the polyp was accommodated in a depression at the broad end of the horn, and the sides of the horn are formed by an external skeletal wall called the epitheca. Many Palaeozoic horn corals seem to have lived partly buried in muddy sediment. Slight overbalancing of the coral as it grew apparently prompted accelerated growth on the side nearest the sea bed to restore the upright attitude, resulting in the curved, horn-like form of the coral. A very different shape is characteristic of 'mushroom corals' such as the scleractinian *Fungia* which is commonly found in the tropics today. Mushroom corals have a relatively flat base and a somewhat convex upper surface where the tentaculate polyp is situated. Compared with horn corals, the shape of mushroom corals makes them more stable on the sea bed. Mushroom corals are peculiar among corals in having the ability to move across the sea bed by executing a series of push-ups. Some extinct tabulate (e.g. *Palaeacis*) and rugose (e.g. *Palaeocyclus*) corals have

## DISTINGUISHING BETWEEN FOSSIL CORAL ORDERS

| | Rugosa | Tabulata | Scleractinia |
|---|---|---|---|
| **Range** | Ordovician–Permian | Ordovician–Permian | Triassic–Recent |
| **Mineralogy** | calcite | calcite | aragonite |
| **Coloniality** | some species | all species | some species |
| **Septa** | bilaterally symmetrical patterning | absent or short | radially symmetrical patterning |
| **Axial structure** | often present | rarely present | often present |
| **Wall pores** | lacking | sometimes present | sometimes present |
| **Epitheca** | present | present | usually lacking |

similar shapes to living mushroom corals and it is possible that they too were mobile.

A wide range of different shapes can be found among colonial corals. In some species the skeletal parts of the individual zooids, known as corallites, are connected only at their bases after which they grow separately to give an organ pipe-like colony. This colony-form is often referred to as fasciculate or phaceloid. It is characteristic, for example, of the tabulate *Syringopora*, the rugosan *Siphodendron*, and the scleractinian *Lophelia*. In other species the corallites remain in close contact with neighbouring corallites throughout their growth. Such tight packing results in the corallites having polygonal cross-sections compared with the roughly circular cross-sections of the separated corallites in fasciculate colonies. These multiserial colonies may adopt a variety of shapes. Some are flat with corallites only present on the upper side of the tabular colony, some have plate-like fronds bearing corallites on both sides, some are roughly dome-shaped, while others have bushy colonies comprising a mass of bifurcating branches. Among the latter is the stagshorn coral *Acropora cervicornis*, a common species on many modern reefs. This a very fast-growing coral, branches extending at more than 10 cm (4 inches) per year, but, compared with dome-shaped coral colonies that often cohabit the same reefs, stagshorn corals are vulnerable to destruction by storms. Many colonies are killed when hurricanes hit and branches are snapped off, some subsequently recovering and recommencing growth in new-found colonies.

Corals are notoriously difficult animals to identify, partly because species may exhibit extreme morphological plasticity in colony form. For example, colonies of the bushy scleractinian *Madracis mirabilis* have widely spaced branches when living in relatively deep water but tightly packed branches in shallower water where currents are stronger. Tightly packed branches slow down current flow through the colony, improving the ability of the polyps to capture planktonic food. Some other reef corals become flattened and plate-like at greater depths as light levels decline. This helps the zooxanthellae to intercept what little light is available for photosynthesis.

Coralliths are the tumbleweeds of the coral world. Approximately spherical in shape these colonial corals grow radially outwards in all directions from a central origin such as a pebble. Unlike most corals, they have no consistent upper and lower surfaces; periodic rolling means that what was once the upward-facing side of the colony, with actively feeding and growing corallites, every so often becomes the dormant side of the colony face down in the sediment of the sea bed. Rolling can be brought about by strong currents, as when storms strike, or by the activities of animals such as fish foraging for food beneath the coral.

New research on modern organisms can often prompt reconsideration of fossils. Such has been the case with the tabulate corals. Tabulates have long been recognised by palaeontologists as a rather mixed group, a 'rag-bag' of Palaeozoic corals that did not fit comfortably within the other cnidarian groups. Up until the 1970s, the chaetetids were generally regarded as tabulates. Chaetetids can be distinguished by having large dome-shaped and lamellar colonies consisting of narrow tubes, the 'corallites', packed tightly together and with a characteristic appearance in cross-section. However, similarities then became evident with an obscure group of modern tropical sponges that have a strikingly similar skeletal structure to Palaeozoic chaetetids. The sponge affinity of chaetetids was subsequently confirmed by the finding of spicules embedded in the skeleton of some Carboniferous species. Chaetetids are now universally recognised by palaeontologists as sponges rather than tabulate corals. None the less, there is still active debate about whether other groups of tabulates are really sponges or corals. Some palaeontologists have argued that an important suborder of tabulates with prismatic corallites – favositids – are sponges. However, the remarkable discovery in a favositid of exceptionally preserved polyps, complete with tentacles, seems to falsify the hypothesis relating them to sponges.

**Above** Thin section of a tabulate coral (larger holes) embedded in a stromatoporoid sponge. About 2 cm in width, this fragment of a 'caunopore' comes from the Silurian of Gotland, Sweden.

One of the most distinctive groups of tabulates are the halysitids or chain corals, especially common in the Silurian. Halysitid corallites are long tubes, oval in section, which in surface views or cross-sections of colonies resemble a series of intertwined chains, each corallite being a link in the chain. It has been shown that some halysitid colonies are in fact formed by several colonies that fused together as they grew. Different parts of the composite colony are traceable to distinct founding corallites (protocorallites) that would have been formed through the settlement and metamorphosis of a separate larva (planula). Another common and easily recognisable group of tabulate corals are the heliolitids, in which the corallites are embedded in a mass of common skeletal material called coenenchyme. These often occur abundantly in reefs of Silurian and Devonian age. Corals participate in symbiotic associations with a multitude of other species in addition to their critical symbiosis with zooxanthellae mentioned earlier. Modern reef corals often associate with fish and crustaceans, both potentially gaining from the protection afforded by the corals' stinging cells. The fossil record seldom furnishes direct and unequivocal evidence of mobile symbionts of corals, but sessile symbionts with hard skeletons can be fossilised with corals. An interesting example is provided by Palaeozoic 'caunopores', a name given to intergrowths between tabulate corals and stromatoporoid sponges. The corals and sponges are so intimately associated that they were originally thought to represent a single species, given the name *Caunopora placanta*. It is now known that the vertical tubes seen in caunopore fossils are colonial tabulate corals belonging to a suborder called the syringoporines, whereas the laminar mass of calcareous material surrounding them consists of a stromatoporoid sponge. Exactly what benefits each symbiont may have accrued from this association is uncertain. However, it is conceivable that the sponge gained protection from predators deterred by the stinging cells of the coral, while the delicate branches of the coral may have benefited from being buttressed by the sponge, thereby enabling it to inhabit environments with strong water currents.

The geological record shows that corals have not always been the main group of organisms building reefs. Early reefs of Cambrian age were built by archaeocyathan sponges, whereas other types of sponges, along with algae and bryozoans, were responsible for most Ordovician reefs. Major reef building by corals commenced in the Silurian. During this period and the succeeding Devonian, tabulate corals, together with stromatoporoid sponges, constructed reefs almost rivalling in size the scleractinian reefs of modern times. Mass extinctions in the late Devonian effectively put paid to coral reefs for the remainder of the Palaeozoic, apart from scattered small build-ups. Carboniferous and Permian reefs were predominantly built by algae, sponges and bryozoans. Scleractinian coral reefs first appeared in the late Triassic. From this time until the present day, tropical scleractinian reefs were widespread, although their geological history has not been continuous; no tropical coral reefs are known from either the early Jurassic or the early Paleocene. These two reef gaps directly follow the terminal mass extinctions of the Triassic and Cretaceous respectively, times of global catastrophe for reef-builders and the inhabitants of reefs.

**Above** Branch, 9 cm long, of the scleractinian coral *Acropora* from the Pleistocene of Yemen.

The late Cretaceous reef-like structures constructed by rudist bivalves eclipse contemporary coral reefs in abundance and importance.

## *ACROPORA*: PALEOCENE–RECENT, WORLDWIDE IN WARM WATERS

The scleractinian coral *Acropora* forms ramose colonies with cylindrical to flattened branches, in some species giving a stagshorn- or elkshorn-like appearance. Corallites generally have 12 septa in two size orders, project from branch surfaces and are very distinct. Unusually among scleractinian corals, polymorphism is developed. There are two types of corallite: long, axial corallites occupy the centres of the branches and form the lead corallites at the branch tips; and smaller radial corallites cover most of the branch surface.

This genus is probably the most common reef-building coral in modern tropical and subtropical seas. Despite the fact that the genus ranges back some 60 million years in geological time, it is only during the last 5 million years that *Acropora* has had this prominent role in reefs.

## *CALCEOLA*: DEVONIAN OF EUROPE, AFRICA AND AUSTRALIA

This unmistakable solitary coral is semicircular in plan view and curved in profile, with the straight face of the semicircle corresponding to the concave side of the curve. Well preserved specimens retain the lid-like operculum, covering the upper surface. The septa are mostly buried by the thick calcareous skeleton that fills much of the interior of the coral.

**Above** The unusual operculum characteristic of the rugose coral *Calceola* is visible in the upper part of this 4 cm high specimen from the German Devonian.

Known as the 'slipper coral', *Calceola* is a Devonian rugose coral distinguished by its hinged operculum. This would have opened to allow the tentacles of the polyp access to the exterior when feeding, and closed to protect the polyp when danger threatened or during periods of dormancy.

## DIBUNOPHYLLUM: CARBONIFEROUS, WORLDWIDE

The solitary rugose coral *Dibunophyllum* is cylindrical to horn-shaped and has an elaborate skeleton best appreciated in cut and polished specimens. As seen in transverse sections, a cobweb-like axial structure at the centre of the coral consists of a long median plate from which diverge septal lamellae that are linked by tabellae. Numerous septa are present, originating around the edge of the corallite and extending inwards almost as far as the axial structure. Dissepiments are small and numerous.

A dark-coloured limestone called the Frosterley Marble, which is not a true marble (i.e. metamorphic), is densely packed with *Dibunophyllum*. The intricate pale grey skeletons of the coral make the polished rock very attractive. Decorative columns of Frosterley Marble can be seen employed in the interior of Durham Cathedral and other mediaeval buildings in northern England.

## FAVOSITES: ORDOVICIAN–DEVONIAN, WORLDWIDE

Colonies of this tabulate coral are usually flat or hemispherical in shape but occasionally have bushy ramose branches. The tightly-packed corallites are small and polygonal, with clearly defined bounding walls. Septa are short, sometimes reduced to spines, and of one size order only. There is no axial structure but the tabulae are well-developed and horizontally disposed. Pores are present in the corallite walls.

**Above** Polished surface of Frosterley Marble sectioning two corallites of the Carboniferous rugose coral *Dibunophyllum*, each about 2.5 cm in diameter.

*Favosites* can be found in many mid-Palaeozoic shallow water deposits, including Silurian reefs where it may co-occur with two other tabulate genera, *Heliolites* and *Halysites*.

**Above** The simple, polygonal corallites of *Favosites* are well seen in this 6 cm high polished block from the Devonian of south-west England.

## *FUNGIA*: PALEOCENE–RECENT OF THE INDO-PACIFIC REGION

Normally solitary, the scleractinian coral *Fungia* has a convex upper surface, a flat or concave underside consisting of an epitheca bearing concentric growth bands, and is approximately circular in plan view. Individual corals can reach diameters of up to about 40 cm (16 inches), but most range from 3 to 15 cm (1–6 inches) in diameter. The upper surface is covered by a large number of radial septa in numerous size orders but there is no axial structure (see p. 25).

Modern examples of the mushroom coral *Fungia* harbour symbiotic zooxanthellae and are restricted to warm, shallow water environments in the tropics and subtropics. They have the ability to move slowly, are capable of righting themselves if overturned and can un-bury themselves if covered by sediment.

## *HALYSITES*: ORDOVICIAN–SILURIAN, WORLDWIDE

*Halysites* colonies have the form of low mounds with the corallites arranged in distinctive chain-like palisades. This gives rise to the common name 'chain coral'. Individual corallites are long, organ pipe-like tubes, circular or oval in cross-section, usually adjoining with two neighbouring corallites except where palisades divide, in which instance they adjoin three other corallites. Septa are absent or weakly developed and are of one size order only. There is no axial structure, and the tabulae are horizontal. Corallite walls lack pores. A single small tubule is present between each adjoining corallite.

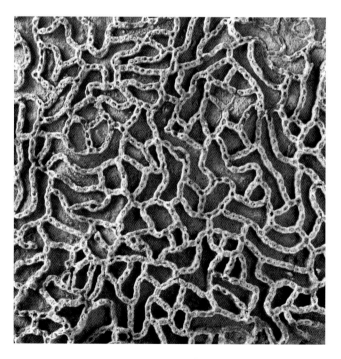

**Above** Part of a colony of the tabulate chain coral *Halysites* from the Silurian of Gotland, Sweden. The visible width of this specimen is 7 cm.

Although often found associated with reefs, *Halysites* can also occur in non-reef settings. An exceptional colony 1.35 m (4.5 feet) in diameter has been reported in the Silurian of Ontario, Canada.

## *HELIOLITES*: ORDOVICIAN–DEVONIAN, WORLDWIDE

Colonies of the tabulate coral *Heliolites* vary in shape from discoidal to dome-shaped or occasionally tree-like. The small corallites are set in a mass of skeletal material called coenosteum (or coenenchyme), which consists of narrow prismatic tubes with cross partitions (diaphragms). Septa are few in number, short and sometimes absent altogether. Horizontal tabulae are present but the corallites lack an axial structure.

*Heliolites* and its relatives constitute a well-defined group of tabulate corals distinguished by the development of coenosteum. This colony-wide skeleton forms a matrix within which the corallites are embedded.

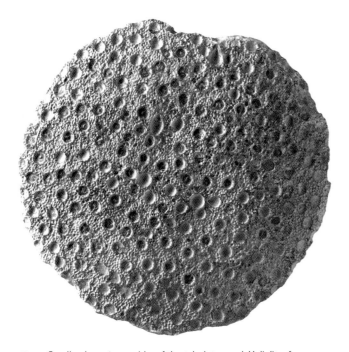

**Above** Small colony, 4 cm wide, of the tabulate coral *Heliolites* from the Silurian of England.

## *ISASTREA*: JURASSIC–CRETACEOUS, WORLDWIDE

*Isastrea* is a colonial scleractinian coral forming encrusting, platey, dome-shaped or sometimes ramose colonies. Corallites are polygonal, 3–15 mm (0.1–0.6 inches) in diameter, with variably developed boundary walls. There are three or four size orders of septa and 30–80 septa in total. Dissepiments are present and sometimes there is a weak axial structure.

Based on comparison with living corals having similar skeletons, it is likely that *Isastrea* had symbiotic zooxanthellae. This may account for its important role in Jurassic and Cretaceous reefs.

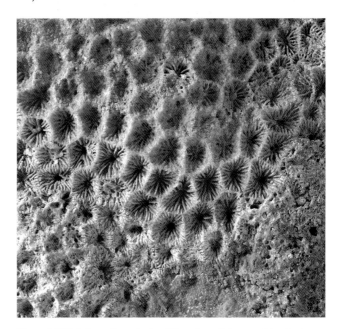

**Above** A British Jurassic example of the scleractinian coral *Isastrea* showing the polygonal corallites, each about 3 mm in diameter and with prominent radial septa.

## *PALAEOCYCLUS*: SILURIAN OF NORTH AMERICA AND EUROPE

This solitary rugose coral is small, circular in plan view with a button-like shape. The upper surface has a regular pattern of radiating septa with dentate edges. Tabulae, dissepiments and axial structures are all absent. The flat underside of the coral is covered by an epitheca with concentric growth lines.

*Palaeocyclus* is particularly well-known from the Swedish island of Gotland where shallow water deposits of Silurian age are magnificently exposed. Individual corals rested freely on the muddy sea bed, much like the extant scleractinian coral *Fungia*, the flat base helping to spread the coral's weight and prevent it from sinking into the mud.

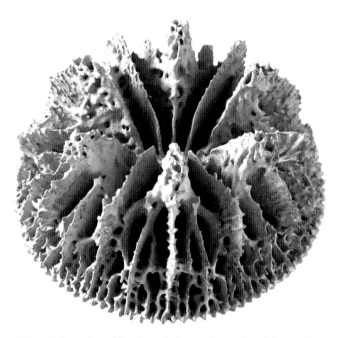

**Above** Oblique view of *Stephanophyllia*, a solitary scleractinian coral 2.5 cm wide, from the Pliocene of Belgium.

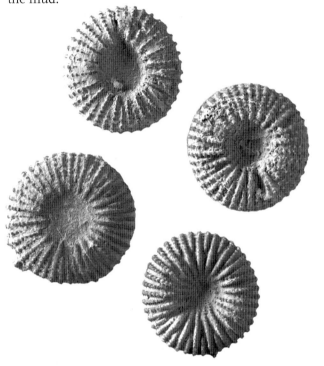

**Above** Four specimens of the free-living rugose coral *Palaeocyclus* from the Silurian of Gotland, Sweden. The button-like specimens are just over 1 cm in diameter.

## STEPHANOPHYLLIA: MIOCENE-RECENT OF EUROPE, AFRICA AND THE INDO-PACIFIC

A free-living, solitary scleractinian coral, *Stephanophyllia* is disc-shaped and usually about 10–35 mm (0.4–1.4 inches) in diameter. The underside of the coral is flat and marked by the radial traces of the bases of the septa. The upperside is convex and bears a regular arrangement of bifurcating septa giving a lace-like appearance. Minute granules cover the faces of the septa. Dissepiments are lacking. An axial columella is present at the centre of the corallite.

While resembling *Fungia* in being discoidal and unattached, living examples of *Stephanophyllia* differ from *Fungia* in not having symbiotic zooxanthellae. Consequently, it is able to inhabit colder, more dimly lit and deeper waters, up to 600 m (2000 feet) in depth.

## STREPTELASMA: ORDOVICIAN–SILURIAN OF NORTH AMERICA, ASIA AND EUROPE

A solitary rugose coral typically cylindrical or horn-shaped, bounded by an epithecal wall. There is a weak axial structure, long major septa and short minor septa that may be lacking. Dissepiments are absent. The tabulae are usually complete, domed but depressed at their centres.

*Streptelasma* is a typical horn coral. It can be found in great abundance in the Ordovician around Cincinnati, USA. Many specimens here show evidence of post-mortem damage due to the activities of boring organisms and abrasion, suggesting that dead corals were exposed for some time on the sea bed before being buried.

**Above** A horn coral, the solitary rugosan *Streptelasma* from the Ordovician of the USA, measuring almost 4 cm in length.

## *STYLOCOENIA*: PALEOCENE–MIOCENE, WORLDWIDE

A scleractinian coral with encrusting, massive or spherical colonies. The corallites are polygonal and small, 2–5 mm (0.1–0.2 inches) in diameter. Six to twelve thin septa are present, in one or two size orders, and a rod-like columella may be present at the axis of the corallite. At one or more corners of each corallite the outer wall is developed into a pillar that projects above the colony surface and has striated sides.

Some colonies of *Stylocoenia* take the form of coralliths. These are ball-shaped, have corallites covering their entire outer surface, and totally envelop the pebble or whatever other substrate the coral initially settled upon. In order to develop, coralliths must be frequently overturned. This can be accomplished by currents sweeping the sea floor or the activities of foraging animals.

## *SYRINGOPORA*: ORDOVICIAN–PERMIAN, WORLDWIDE

The mound-shaped colonies of the tabulate coral *Syringopora* (see colour fig. 3) comprise numerous narrow, cylindrical corallites that multiply by branching and are cross linked by small tubules. Corallite walls are thick and non-porous. Septa are greatly reduced, usually represented by longitudinal rows of spines or absent altogether. The tabulae sag downwards and are sometimes drawn together to form an axial tube at the centre of the corallite.

*Syringopora* is among the most abundant and long-lasting of all Palaeozoic coral genera, and is particularly common in limestones of Carboniferous age. Colony organisation in this coral is classified as 'phaceloid'; unlike many other colonial corals, the corallites are not packed tightly together, and during life there would have been no connections between the soft tissues of the mature polyps located atop the cylindrical corallites which are spatially separated from one another. This suggests a low level of colonial integration, the individual zooids functioning individually.

**Above** Spherical colony of the scleractinian coral *Stylocoenia*. This 'corallith' from the French Eocene has a diameter of 1.5 cm.

# BRYOZOANS

Although bryozoans are abundant today and have a rich fossil record, to the non-specialist they are one of the most unfamiliar of all invertebrate groups. More than 6000 living species exist in the phylum Bryozoa. However, bryozoans have no widely used or consistent common name; they are occasionally referred to as moss animals, sea mats or lace corals. Colonies of bryozoans at first sight can resemble plants (hence the name 'moss animals'), many grow as flat encrustations on seaweeds or sea shells ('sea mats'), or form meshworks looking like a delicate coral ('lace corals').

None of these vernacular names used for bryozoans give any clue to the true biological affinities of the phylum. In fact bryozoans belong to a group of invertebrates called lophotrochozoans. This group includes annelids, molluscs and also brachiopods to which bryozoans are traditionally allied. Like brachiopods, bryozoans feed on plankton by means of a lophophore, and have bodies containing fluid-filled coelomic cavities, representing a more advanced anatomy than that found in corals and sponges. In most other respects, however, there seems to be little similarity between bryozoans and brachiopods. Bryozoans are entirely colonial animals, with colonies consisting of a few to tens of thousands of genetically identical modules termed zooids, whereas brachiopods are solitary animals. Individual zooids tend to be less than 1 mm (0.04 inches) in size compared with the centimetric size of brachiopods. The bryozoan lophophore (see colour fig. 2) is an inverted cone- or horseshoe-shaped ring of tentacles, contrasting with the more complex lophophores of brachiopods with their very numerous filaments. Tiny hair-like cilia arranged along the tentacles beat in unison to create a flow of water that propels planktonic food particles towards the mouth, which lies at the centre of the cone. When feeding, the individual zooids function like miniature suction pumps. Unlike the writhing tentacles seen in corals and anemones, the tentacles of bryozoans can be held still when feeding, occasionally flicking to bat food particles in the direction of the mouth or to expel unwanted particles out of the lophophore. The lophophore is able to withdraw into the safety of the zooidal walls if danger threatens, for example from a predatory sea-spider. Withdrawal is accomplished by the lightning-fast contraction of a retractor muscle. Bryozoans have U-shaped guts, the anus being outside the lophophore, in contrast with the bryozoan-like phylum Entoprocta where the anus opens within the lophophore.

Most bryozoan colonies begin life with the settlement of a tiny swimming larva, formed as a result of sexual reproduction, on a hard surface such as a rock or shell. The larva undergoes a catastrophic metamorphosis in which all of the cells are radically rearranged and new tissues begin to form. This results in the first zooid of the new colony which is called the ancestrula. All subsequent zooids are produced by an asexual process of budding involving repeated division of the expanding body into modular units – zooids – of more or less identical shape and size. As with other colonial animals, the zooids within a colony are genetically identical, that is they are clones of one another. Zooids can be likened to building blocks; the size of the colony depends mostly on the number of blocks (zooids) present, while its shape is a consequence of how they are arranged relative to one another. Just as architecturally varied buildings can be fabricated from identical blocks laid in different ways, so bryozoan colonies of diverse shapes can be constructed by identical zooids budded in different configurations.

While it is true that some bryozoan colonies are indeed made up of identical zooids, colonies of most species actually contain two or more distinct kinds of zooids, known as polymorphs, serving different functions. Autozooids are the basic feeding polymorphs present in all colonies. These can be accompanied by polymorphs used for reproduction, defence, support or cleaning the colony surface, as well as others whose role seems to be simply to fill the spaces between the autozooids so that their lophophores do not overlap. Any polymorphs that are incapable of feeding must be nourished by those that can feed. Nourishment is made possible by soft tissues that interconnect the zooids within a colony, passing through pores in the skeletal walls or draping over the colony surface. Not only can food be exchanged between zooids, but nervous impulses may also be transmitted through the colony. For example, touching the lophophore of one zooid can result in the retraction of all of the zooids across a wide area of the colony.

Probably the most remarkable bryozoan polymorphs are the avicularia, named for their occasional resemblance with a bird's head. Avicularia have beaks or mandibles that are capable of grasping animals, including sea spiders and small worms, attempting to prey on the bryozoan. They can hold the predator tightly until it dies. The mandibles of avicularia evolved through the modification of trapdoor-like opercula that close the orifice of autozooids when the tentacles are retracted. Because avicularia themselves are unable to feed, the energy-giving resources they need to operate the powerful muscle closing the mandible must be supplied by the feeding zooids in the colony. The trade-off is that the autozooids and indeed the colony as a whole may benefit from the activities of the avicularia in discouraging predators.

## BRYOZOAN CLASSIFICATION

Class Phylactolaemata (Permian–Recent)

Class Gymnolaemata
    Order Ctenostomata (Ordovician–Recent)
    Order Cheilostomata (Jurassic–Recent)

Class Stenolaemata
    Order Cyclostomata (Ordovician–Recent)
    Order Cystoporata (Ordovician–Triassic)
    Order Trepostomata (Ordovician–Triassic)
    Order Cryptostomata (Ordovician–Triassic)
    Order Fenestrata (Ordovician–Permian)

Avicularia are ineffective in defence against large predators such as fish. Being unable to escape by moving away, bryozoans can only defend themselves against large predators by being poisonous. Bryozoan poisons include alkaloids and a unique set of organic compounds called bryostatins which have attracted pharmaceutical interest because of their ability to reduce cancerous tumours. It has recently been shown that bryostatin in the bryozoan *Bugula neritina* is actually manufactured by a bacterium that lives symbiotically within the tissues of the host animal. The presence of bryostatin makes the larvae of *Bugula neritina* distasteful to fish.

One small group of bryozoans – the phylactolaemates – live exclusively in freshwater. They lack mineralised skeletons and are consequently unimportant as fossils, although their seed-like dispersal structures (statoblasts) have been found in freshwater sedimentary rocks dating back to the Permian. Marine bryozoans, in contrast, are very well-represented in the fossil record as a result of the calcite or aragonite skeletons present in the majority of species. The taxonomy of living bryozoan species is based largely on features of these skeletons. As these features are preserved in fossil bryozoans, there has been no need for palaeontologists and zoologists to develop different systems of identification and classification. The same is not true, however, of all groups of invertebrates with fossil representatives (e.g. serpulid worms; see Chapter Four).

Unusually for a phylum with a good fossil record, bryozoans have yet to be found in the Cambrian; their known fossil history begins in the early Ordovician when the phylum underwent a spectacular evolutionary radiation. Particularly abundant in the Ordovician are trepostomes, so-called 'stoney bryozoans', which range through into the Triassic. Many species of trepostomes have colonies resembling small bushes, while others are dome-shaped and a few are frondose. The colony surface may be covered by tiny hummocks – monticules – which represent places where water filtered of food particles by the zooids was channelled before being jetted away from the colony surface. Similar monticules are present in some present-day bryozoans and have been observed to function in exactly this way. The zooids of trepostomes take the form of long curved tubes, polygonal in cross-section and often divided by numerous cross-partitions called diaphragms. Growth of the tubular zooid was accompanied by the formation of new diaphragms, the outermost of which formed the floor of the living chamber of the tentaculate zooid, gradually 'jacked-up' as the zooid increased in length. Small polymorphs (mesozooids and exilazooids) may be present between the autozooids in trepostome colonies but their function is unclear. Tree-like trepostome colonies are seldom preserved intact. Instead, they are usually fragmented into a series of branches. The major constituent of some Ordovician limestones comprises branches of trepostome bryozoans, as in the Cincinnati region in the mid-west of the USA.

Another common group of fossil bryozoans, the fenestrates, became increasingly important in the late Palaeozoic before meeting their demise during the mass extinction at the end of the Permian. Fenestrates generally have more delicate colonies than trepostomes, often net- or mesh-like in form, with narrow branches linked at intervals, either by solid rods of skeleton called dissepiments or cross branches. Only one side of the branches bears apertures through which the lophophores of the feeding zooids would have protruded during life. After filtering out food particles, expended water was probably expelled through the gaps between branches and dissepiments and away from the colony, as in some recent bryozoans with similar shaped colonies. The efficient one-way flow of colonial water currents in

**Below** Screw-like central axes of the Carboniferous fenestrate bryozoan *Archimedes*. The field of view is a little under 10 cm in width.

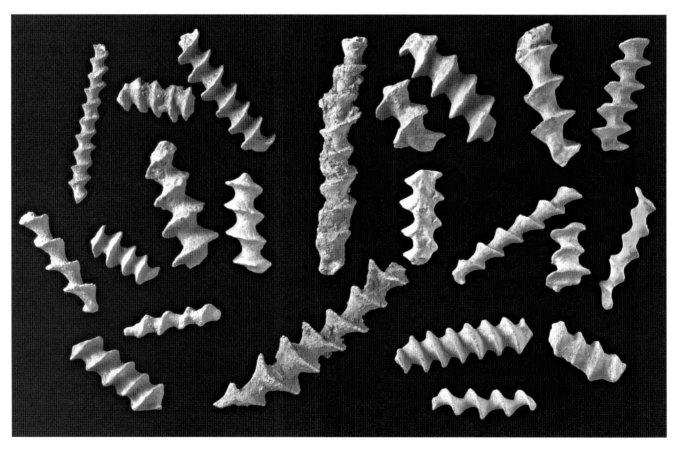

fenestrates was very different from that based on monticules present in trepostomes. Some Carboniferous and Permian fenestrates inhabited reef-like mud mounds. It has been hypothesised that these bryozoans formed sediment baffles which, along with the zooids defecating pellets of sediment, caused fine-grained sediment to be deposited around the bases of the bryozoan colonies and contributed to the formation of these mounds.

The most distinctive of all fossil bryozoans is a fenestrate genus named *Archimedes* after its resemblance to a water pump – the Archimedes Screw – invented by the Greek philosopher Archimedes of 'Eureka!' fame. The screw-like fossil is in fact merely the central axis of the colony; a typical fenestrate meshwork is attached to the edges of the screw in intact colonies but in most instances is broken off. Colonies of *Archimedes* were able to grow very tall – more than 1 metre in height – but were prone to toppling over. A new screw axis could then develop on the side of the prostrate old axis, restoring upward growth of the bryozoan.

Cyclostomes and cheilostomes are the principal bryozoan groups found as fossils in Mesozoic and Cenozoic rocks. They provide a fascinating comparison between biological groups with different evolutionary origins and biological attributes that have shared the same environments for more than 150 million years of geological time. Whereas cyclostomes originated in the Ordovician, cheilostomes are relative newcomers having first evolved in the Jurassic. Present day cheilostomes are usually superior to cyclostomes, notably in feeding and competition for living space. This superiority seems to have been established at the outset of their coexistence. It is possible to observe the results of competition for living space between cheilostomes and cyclostomes in fossils by looking for instances in which one encrusting colony overgrows the edge of another and smothers the underlying zooids. When such 'frozen in time' overgrowths between fossil cheilostomes and cyclostomes are studied it is found that cheilostomes are

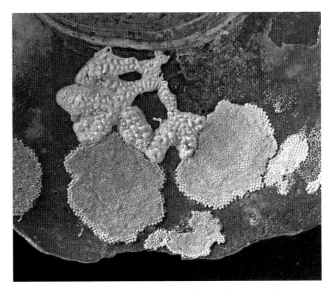

**Above** Several Pliocene bryozoan colonies 'frozen' in competition for living space on the surface of a bivalve shell from New Zealand. Field of view is 3.5 cm wide.

the winners in about 65–70 per cent of cases. None the less, cyclostomes have continued to exist in the face of this competition from the dominant cheilostomes, although their abundance may have declined as a result of competition.

One solution to the problem of the constant threat of being overgrown, both by other bryozoan colonies and the myriad organisms that encrust hard surfaces, is for colonies to adopt a fugitive lifestyle. Rather than resisting overgrowth, fugitive colonies attempt to disperse their zooids as widely as possible over the surface they encrust in the hope that at least some of these zooids will escape overgrowth. Such wide dispersal is best achieved by having a branching colony form with long zooids and a fast growth rate. A fossil cyclostome – *Stomatopora* – common in the Jurassic and Cretaceous has this colony form, as does the cheilostome *Herpetopora* which can be found encrusting bivalves and echinoids in the Cretaceous Chalk of Europe.

Bryozoan colonies that grow erect may also avoid the danger of overgrowth posed by competitors for substrate space. However, they do so at the expense of facing a different problem, that of dislodgement or breakage by strong water currents followed by transportation into unfavourable environments where survival is unlikely. One way of minimising this risk is to have a flexible colony that literally 'goes with the flow', bending to absorb the energy of currents. The living cheilostome bryozoan *Flustra* exhibits this strategy well: the frondose colony has a very weakly mineralised skeleton that can flex in strong currents. Even so, vast numbers of *Flustra* colonies are routinely ripped up from the sea bed and cast ashore on beaches around the UK following storms. They can be found among the flotsam and jetsam, looking like the seaweed implied by their common name, horn-wrack. Not surprisingly in view of its feebly mineralised skeleton *Flustra* has no fossil record.

In marked contrast to *Flustra*, a distinctive bryozoan called *Terebellaria* illustrates an alternative strategy – skeletal strengthening – for coping with currents. The bushy colonies of this Jurassic cyclostome reinforce their basal branches, where stresses are greatest, by covering them with multiple layers of overgrowing zooids. Originating from the tips of the branches, the overgrowing layers advance downwards towards the colony base, smothering the older zooids of earlier layers while thickening the colony such that branches near the base are thickest and strongest. Usually the leading edge of the overgrowth is in the form of a continuous spiral wrapped around the branches, but even within the same colony it can alternatively comprise a series of discrete rims. Isolated branches of *Terebellaria* resemble tiny fairground helter-skelters. Colonies of *Terebellaria* are common in the mid Jurassic rocks forming the cliffs behind the D-Day beaches of Normandy, France.

Adaptation to an extreme mode of life can be seen in the cheilostome bryozoan *Selenaria*. The cap-like colonies of this Australasian genus are about 1 centimetre in

**Above** *Terebellaria*, an unusual cyclostome bryozoan from the Jurassic of Normandy, France. This example is 1.8 cm tall.

diameter, circular in plan view, with all of the zooids opening on the upper convex surface of the colony. Remarkably for a bryozoan, colonies of *Selenaria* are able to move across the sea bed through their own actions (see colour fig. 5). This is made possible using the appendages belonging to some polymorphic zooids called vibracula. In living specimens these stilt-like appendages, called setae, can be seen to project downwards from the outer edge of the colony and support it a little way above the sediment surface. Co-ordinated movements of the setae cause the colony slowly to lunge around. Why they move is uncertain, but this capacity does enable colonies buried by sudden influxes of sediment to dig their way to the surface. Fossil examples of *Selenaria* can be found, although they are not as common as some other 'lunulite' genera (e.g. *Lunulites* and *Discoporella*) which have an almost identical colony form but lack the capacity to move. The evolution of adaptations permitting the mobility of *Selenaria* can be traced in the fossil record. Ancestors of lunulites were conventional encrusting bryozoans with polymorphic zooids – the avicularia mentioned above – with mandibles potentially functioning in defence. In the late Cretaceous some of these ancestral species evolved the ability to grow beyond the margins of the substrates they encrusted. Colonies of this type no longer required a large substrate and instead could develop on sand grains and other tiny substrates, effectively becoming free-living. Such colonies evolved a more regular shape than encrusting colonies as they were able to grow without hindrance from obstacles on the substrate. The defensive mandibles of the avicularia increased in length to become the colony-supporting setae of the vibracula. Not until the Miocene, however, did mobility evolve, and this happened only in the genus *Selenaria*. For mobility to occur a ball and socket joint evolved that permitted the setae to move in

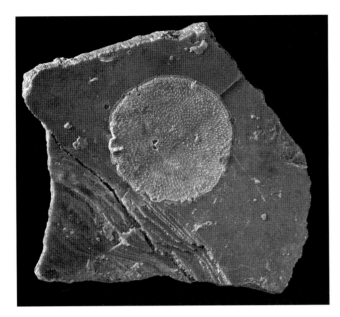

**Above** Colony of the disc-like cyclostome bryozoan 'Berenicea', 1.7 cm in diameter, encrusting a shell fragment from the Jurassic.

all three dimensions, in contrast to the movements of setae in other lunulites that are restricted to one plane.

Bryozoan colonies form habitats that are exploited by many other animals: crustaceans and worms often make their homes among the branches of bryozoans, while small fish may use them as places to shelter. Although these associates are seldom fossilised with the bryozoans, other animals that become intergrown with their bryozoan hosts are fossilised. Branches of the cheilostome bryozoan *Celleporaria* found in the Miocene and Pliocene of Europe often contain individuals of the small scleratinian coral *Culicia*. Although the corals are sometimes found fully preserved, dissolution of the aragonitic coral skeleton during diagenesis often leaves characteristic holes in the surface of the calcitic skeleton of the bryozoan where the corals were formerly located.

## 'BERENICEA': TRIASSIC–RECENT, WORLDWIDE

'Berenicea' colonies occur as thin circular to irregular patches encrusting shells and rocks. Newly-budded zooids occur around the circumference of the colony where a thin lamina may extend across the substrate out beyond the budding zone. Autozooids have slightly convex, minutely porous outer frontal walls. A circular or oval aperture is located at the end of the frontal wall. In well-preserved specimens a tubular peristome oriented perpendicular to the colony surface encloses the aperture. Larval brooding polymorphs – gonozooids – are larger and more bulbous than the autozooids. These vary in shape, with different species having gonozooids with circular, sac-shaped, triangular and crescent-shaped outlines.

One of the most common bryozoans in the Mesozoic, particularly the Jurassic, is the cyclostome 'Berenicea' which forms small patch-like colonies encrusting shells and rocks. Strictly speaking the name *Berenicea* has no validity as the species on which it was founded in 1821 cannot be recognised. However, the name written in quotes is a useful shorthand for an assortment of species now formally placed in genera such as *Plagioecia*, *Mesonopora* and *Hyporosopora*. Most of these genera are distinguished by having different shaped gonozooids. Unfortunately, not all colonies develop gonozooids, leaving a residue of specimens that can be difficult or impossible to place in a genus.

## CHASMATOPORA: ORDOVICIAN–SILURIAN OF EUROPE, ASIA AND NORTH AMERICA

Colonies of *Chasmatopora* (see colour fig. 4) are mesh-like, constructed of branches that bifurcate and anastomose in a single plane. The holes (fenestrules) are elongated parallel to the growth direction of the colony. Frontal sides of branches on one face of the meshwork bear 2–8 rows of circular zooidal apertures, their reverse sides lack apertures but are faintly striated. There are no polymorphic zooids.

The mesh-like colony of *Chasmatopora* typifies most bryozoan genera belonging to the order Fenestrata. It should be noted that colonies of very similar appearance subsequently evolved independently in two other bryozoan orders, Cheilostomata and Cyclostomata. In life a tentacle crown would have emerged from each aperture, pulling water towards the frontal side of the colony and expelling water filtered of plankton through the holes and away.

## DISCOPORELLA: MIOCENE–RECENT, WORLDWIDE

The free-living colonies of *Discoporella* are small, regular and cap-shaped, circular in plan view with a convex upper surface and concave or flat underside. Zooids open on the upper surface only; the lower surface comprises the basal sides of the zooids and is radially sectored and porous. Autozooids are diamond- or rhombic-shaped, with depressed frontal walls pierced by numerous holes (opesiules) for the passage of muscles. Immediately beyond the oval orifice of each autozooid is a small avicularium.

The skeleton of this 'lunulite' anascan cheilostome and the closely related genus *Cupuladria* are made entirely of aragonite. This mineral is slightly more dense than calcite, probably making these free-living colonies heavier and somewhat less likely to be overturned by currents sweeping the sea bed.

**Above** Variously oriented, fossil specimens of *Discoporella*, a free-living cheilostome bryozoan. Each colony is about 3 mm in diameter.

## 'FENESTELLA': ORDOVICIAN–PERMIAN, WORLDWIDE

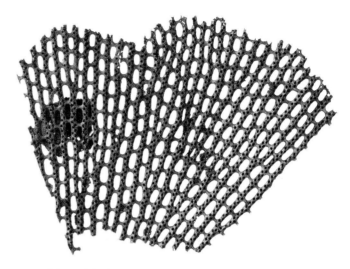

**Above** Colony of the fenestrate bryozoan '*Fenestella*' about 2.1 cm wide. Silicified after burial, the meshwork was extracted by acid dissolution of limestone from the Carboniferous of Northern Ireland.

Mesh-like colonies of '*Fenestella*' consist of bifurcating branches laterally inter-linked by dissepiments. The frontal surfaces of the branches bear two rows of circular zooidal apertures separated by a ridge that is sometimes ornamented by nodes. There are no apertures on the dissepiments that are narrower than the branches they connect. The holes or fenestrules in the meshwork are usually rectangular and elongated parallel to growth direction. Outgrowths of skeleton in the form of barbed spines are sometimes present, particularly close to the colony base.

One of the most common of all bryozoan genera, '*Fenestella*', has recently been subdivided into a series of new genera that are distinguished mainly on the basis of differences in the three-dimensional shape of the zooidal chambers, as reconstructed from thin sections. Therefore, the genus name is here written in quotes to indicate its use in the broader sense. Colonies fed like those of the related fenestrate genus *Chasmatopora* by generating a one-way flow of water through the fenestrules, filtering food particles in the process.

## *FISTULIPORA*: SILURIAN–PERMIAN, WORLDWIDE

Encrusting, mound-like and thickly branched colonies can all be found in the cystoporate bryozoan genus *Fistulipora*. Zooidal apertures are roughly circular but in some species are deformed by a crescentic edge of smaller radius on one side (lunarium). Between the apertures are areas of cyst-like vesicles best seen in vertical sections. The surfaces of larger colonies may be festooned by regularly spaced patches (maculae) devoid of zooidal apertures and often raised as low mound-like monticules.

As colonies of *Fistulipora* grew larger and older zooids became more and more distant from the colony margin, it was no longer possible to dispose of all filtered water at the outer margin of the colony. The solution to this problem was to develop maculae, patches lacking feeding zooids where filtered water could be channelled for expulsion. Maculae often incorporate a hummock, in which case they are termed monticules. They can be found in a great variety of different bryozoans and provide an example of parallel evolution of a feature of clear functional importance to the living bryozoan colony.

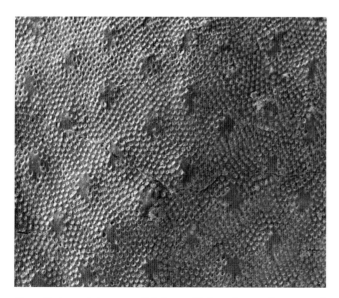

**Above** Cystoporate bryozoan *Fistulipora* with the colony surface covered by dimple-like monticules. The field of view in this is 3 cm wide.

## *HALLOPORA*: ORDOVICIAN–SILURIAN, WORLDWIDE

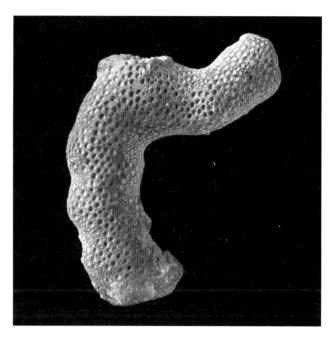

**Above** Branch, 15 mm high, of the trepostome bryozoan *Hallopora*, a genus particularly common in the Silurian which had bushy colonies when alive.

Colonies of *Hallopora* grew as small bushes but are invariably found broken up into branch fragments. Branches are cylindrical in form, a few millimetres in diameter, and bifurcate at intervals. The apertures of autozooids interspersed with polymorphs called mesozooids cover the branch surfaces. Autozooidal apertures are larger and more rounded than those of the mesozooids that surround them. Sectioned specimens show the long, tubular internal shape of the zooids, with greater numbers of cross-partitions (diaphragms) being present in the mesozooids than the autozooids.

*Hallopora* is a typical trepostome bryozoan, a group that can be very abundant in Ordovician and Silurian fossil assemblages, such as the Cincinnatian of Ohio, Indiana and Kentucky. While not strictly reef-builders, these bushy trepostomes would have formed thickets on the sea bed, providing habitats and shelter for a range of other animals.

## *HERPETOPORA:* CRETACEOUS OF EUROPE, ASIA AND NORTH AMERICA

**Above** The runner-like cheilostome bryozoan *Herpetopora* spreading out across the surface of a bivalve mollusc shell from the Cretaceous Chalk of England. Each thread-like zooid is roughly a millimetre long.

This primitive cheilostome has encrusting colonies that are typically found on bivalve shells or echinoid tests. The zooids are arranged in single rows to form slender branches, new branches originating from the sides of the existing branches, often in pairs with one on the left and the other on the right side. Autozooids are long and slender, initially very narrow but broadening towards the large oval aperture that is sometimes permanently sealed by a cover called a closure plate. Unlike most cheilostomes, *Herpetopora* has neither ovicells nor avicularia. Non-feeding polymorphs – kenozooids – with reduced apertures are present and vary in shape; some are extremely long and filament-like.

*Herpetopora* is an extreme example of a bryozoan runner. Colonies probably grew rapidly, dispersing their zooids very widely across the substrate. The skeleton often shows evidence of repaired damage, showing that colonies were able to survive local destruction of zooids.

Zooidal lophophores can be inferred to have been distantly spaced, in contrast to most bryozoans where the tentacles are tightly packed.

## *HEXAGONELLA:* PERMIAN OF ASIA AND AUSTRALIA

This cystoporate bryozoan has flattened, erect branches with an internal wall (median lamina) at their centres. Low ridges on the colony surface mark out the boundaries of irregular polygonal subcolonies. Zooids are absent from the centres of these subcolonies but their circular apertures are arranged in radiating rows towards the edges of the subcolonies. The planar, slightly pustulose colony surface between the apertures can be seen to comprise a vesicular skeleton in sectioned specimens.

**Above** A 5 cm long branch of *Hexagonella* from the Permian of Australia. This cystoporate bryozoan has polygonal subcolonies bounded by ridges.

In many bryozoans, the colony is divisible into a series of subcolonies, each comprising a cluster of zooids more clearly associated with one another than they are to the other zooids of the colony. The distinctiveness of these subcolonies varies: in some bryozoans, it is very difficult to define the precise boundaries between subcolonies, whereas in others the boundaries are unequivocal. *Hexagonella* falls into the latter category as ridges on the colony surface clearly mark out the subcolonies.

## *MELICERITITES*: CRETACEOUS OF EUROPE AND ASIA

The bushy colonies of *Meliceritites* have bifurcating branches that are narrow and circular in cross-section. Autozooids are long and horn-shaped when seen in vertical sections of branches. On the branch exterior they have a polygonal outline shape, with the frontal surface covered by a minutely porous wall containing an aperture in the form of a half ellipse. The straight edge of the aperture is the hinge line of a calcified lid (operculum), occasionally preserved *in situ*, which sealed the aperture when the tentacles were retracted. Polymorphic zooids include brooding gonozooids, bulbous and oval in outline shape, and eleozooids in which the aperture and operculum is enlarged relative to the frontal wall of the zooid. Eleozooids may be smaller, the same size as or larger than autozooids, and their apertures can have rounded or pointed ends.

While the colony form of the cyclostome bryozoan *Meliceritites* is unremarkable, recurring in numerous different groups of bryozoans, the operculum and eleozooids are noteworthy. Normally lacking in cyclostomes, opercula are a feature more typical of cheilostome bryozoans. Their occurrence in melicerititid cyclostomes is an extraordinary example of parallel evolution. Moreover, melicerititid eleozooids with their enlarged opercula resemble cheilostome avicularia and probably had a similar role in defending the colony against small predators.

**Above** Branching colony of the cyclostome bryozoan *Meliceritites*, a little less than 4 cm high, from the Cretaceous of southern England.

## *MICROPORELLA*: MIOCENE–RECENT, WORLDWIDE

Sheet-like encrusting colonies typify the ascophoran cheilostome *Microporella* although a few species have erect foliaceous colonies. The autozooids are hexagonal in outline shape and heavily calcified, with a convex frontal wall bearing pustules and perforated by small pores (often blocked by sediment or cement in fossils). The orifice through which the tentacles emerge is a half ellipse with a straight lower edge. Just beneath the orifice is a large pore, the ascopore, sometimes circular but more often kidney-shaped, and with raised edges. Small pointed avicularia occur on the frontal wall of the autozooids to the side of and just below the orifice, singly or in pairs. Hood-like ovicells are present in some autozooids.

The ascopore of *Microporella* is an important structure in the functioning of the zooids. When the lophophore is protruded, sea water passes through the ascopore and

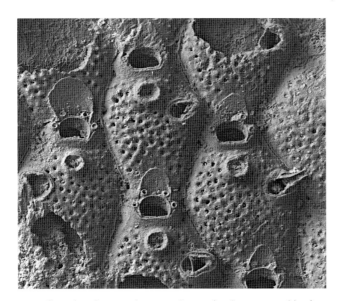

**Above** Scanning electron microscope image showing some zooids of the cheilostome bryozoan *Microporella*, each about 0.5 mm long, from the Miocene of New Zealand.

into a sac called the ascus beneath the calcified frontal wall in order to compensate for the volume of space vacated by the lophophore. Retraction of the lophophore forces the ascus to empty, sea water flowing out through the ascopore. Encrusting ascophoran cheilostomes such as *Microporella* are extremely diverse, both today and in Cenozoic fossil assemblages. Hidden surfaces, including the undersides of corals, shells and rocks, are usually the most favoured habitats. Here the colonies are less likely to fall victim to predators, burial by sediment, or overgrowth by algae which are excluded by the low light levels.

## *PTILODICTYA*: ORDOVICIAN–DEVONIAN, WORLDWIDE

This distinctive cryptostome bryozoan has a straight to gently curved frond which, unusually for a bryozoan, does not bifurcate. Well-preserved specimens preserve a conical point at the lower end of the frond. In cross-section the frond is flattened and oval to almost diamond-shaped, containing a central wall (lamina) along which specimens sometimes split. Occasionally low mounds – monticules – are present on the frond

surface. The rectangular zooidal apertures are arranged in well-defined rows along the length of the frond; in some specimens there is a central zone of parallel rows flanked by outer zones in which the rows diverge slightly from the frond axis.

During life the conical end of the frond of *Ptilodictya* fitted via an articulated joint into the crater of a volcano-shaped base that was firmly cemented onto shells and rocks. Judging from living bryozoans with similar joints, the frond would have been able to sway slightly in the current.

**Above** Frond of *Ptilodictya* measuring 4 cm long. Collected from Silurian rocks, this cryptostome bryozoan was articulated at the base when alive.

## *STEGINOPORELLA*: EOCENE–RECENT, WORLDWIDE

Most species of *Steginoporella* form sheet-like encrusting colonies although some grow erect as hollow branches or fronds. The zooids are large, box-shaped and hexagonal in surface outline. The upper surface comprises a depressed frontal wall (cryptocysts) in which a large aperture is situated. Indentations in the bottom corners of the aperture were for the passage of muscles during life. Often, colonies have polymorphic zooids similar in shape to the predominant autozooids but with slightly larger apertures.

The anascan cheilostome *Steginoporella* lives mostly in warm water environments today. It is of particular interest because of the presence in some species of polymorphic zooids that are in some ways intermediate between autozooids and avicularia; they are able to feed, as in normal autozooids, but have enlarged mandibles like avicularia.

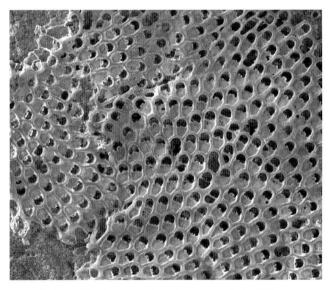

**Above** Each zooid is about 1 mm long in this New Zealand Pliocene specimen of the cheilostome bryozoan *Steginoporella*.

## SPONGES

Sponges are known to many people only as useful objects in the bathroom or for washing their cars. These 'bath sponges' belong to a phylum of primitive animals called the Porifera that has a fossil record extending back some 600 million years to the late Precambrian. They occupy the basal rung of the evolutionary tree of multicellular animals. More than 10,000 species of sponges are living today. All are aquatic, most being marine but some inhabiting freshwater environments.

Sponges are the simplest of multicellular animals; they lack a mouth and have no distinct organs. Remarkably, it has been shown for some species that if the animal is passed through a sieve to separate the individual cells, these cells will gradually come back together over an interval of several hours to reconstitute the multicellular animal. This ability is unique to sponges among animals and plants, and shows how cells from the same individual are able to recognise one another and reassemble in the correct configuration even in such primitive animals. Given the liberal behaviour of individual sponge cells, it is not surprising that sponge animals show a great plasticity in shape and are perfectly able to survive after pieces are broken off the body (partial mortality). In one group of sponges (hexactinellids or glass sponges), the cells of the body are actually fused to create a huge network containing many nuclei called a syncytium.

Plasticity in growth and partial mortality make it appropriate to treat sponges as colonial animals from an ecological standpoint, even though they do not grow by budding differentiated modular zooids like those of bryozoans or colonial corals. However, a type of modularity is developed in some sponges as a result of the occurrence of multiple water outlets, as described below.

Sponges are active suspension feeders (filter feeders); they pump water through their bodies, removing particles of planktonic food in the process. Pumping is accomplished using specialised cells called choanocytes, each equipped with a hair-like flagellum. Passing food particles adhere to the sticky surfaces of the choanocytes and can be digested. Prodigious rates of water pumping have been measured in living sponges. Some fist-sized sponges are able to pump 50 litres of water per hour, while others can filter up to 20,000 times their body volume of water each day. Bacteria and dinoflagellate algae are the most important components of the sponge diet, but dissolved organic matter in the water can also be utilised. A species of sponge living in caves in the Mediterranean Sea has recently been discovered that, unusually, eats small crustaceans, trapping them in spicules on the outside of the body.

The most basic body shape found in sponges is a barrel-like cylinder enclosing a cavity known as a spongocoel. A large opening at the top of the spongocoel is called the osculum. In addition, there are numerous smaller openings along the sides of the body called ostia. During feeding, water enters into the sponge through the many ostia, food particles are extracted, and waste currents are passed into the spongocoel from where they exit through the osculum. The surface area of the osculum is normally less than the combined surface areas of the ostia, with the result that the exhalent water is confined to a narrower outlet and therefore leaves the sponge at a higher velocity than the inhalent water enters. This has the effect of producing a powerful jet taking filtered water well away from the sponge, thereby minimising the chances that it will be re-filtered. More complex sponges have several oscula, each forming a separate venting point for exhalent currents. This imparts modularity in structure and function, similar to that found in true colonial animals. The oscula in these modular sponges are often situated atop finger-like branches.

Recent sponges tend to be vivid yellow, orange, red or blue in colour. Encrusting species may resemble brightly coloured pieces of plastic draped over rocks and shells on the sea bed. Many other sponges grow upright and have cylindrical (see colour fig. 6) or cup-shaped bodies. Coral reefs are home to diverse assemblages of sponges varying in size and shape. Giant sponges several metres in height may occur, forming important habitats for fish and other animals. Even the ocean depths are colonised by sponges, particularly the lattice-like glass sponges that mostly live at depths between 200 and 2000 metres. Sponges manufacture a vast array of toxins and other chemicals to assist their survival against predators, competitors and diseases. The possible use of these natural products in pharmacology has stimulated considerable research on the biochemistry of sponges.

This idea is not new; the Maori people of New Zealand have long used the 'crumb of bread sponge' *Halichondria moorei* to assist the healing of wounds.

An important characteristic of sponges is their ability to secrete tiny, needle-like skeletal elements called spicules. These spicules vary considerably in details of their shape between different species of sponges, making them of great value in sponge taxonomy. Spicule shape is often a far better guide to the generic or specific identity of a sponge than is the overall shape of the body within which the spicules are embedded. Some spicules have the form of straight needles tapered at both ends, others consist of three or more rays radiating from a central point, precisely 'engineered' in a star-like configuration. Tiny spikes may adorn the spicule surface, and some spicules have a bulb-like rounded end. Two or more types of spicule can occur together in a single sponge individual. These often fall into two size classes; small (microscleres) and large (megascleres). Chemically, sponge spicules can be made of spongin, which consists mostly of the protein collagen, the calcium carbonate mineral calcite, or hydrated silica similar to opal in composition. Bath sponges have spicules of spongin, whereas the sponges found in the fossil record typically have calcareous or siliceous spicules, explaining the conspicuous contrast in hardness between sponges of the domestic and palaeontological realms. Some present-day and fossil sponges lack spicules altogether. Mineralised sponge spicules are routinely fossilised. However, they can become widely dispersed following decay and disaggregation of the soft parts of the sponge. Dispersed sponge spicules are occasionally present in sufficient abundance to form a sedimentary rock called spiculite. Sponges with fused or strongly interlocking spicules are more likely to survive intact as recognisable macrofossils. In addition to the spicular skeleton, some sponges have a solid basal skeleton of calcium carbonate that also fossilises well.

Palaeontologists are inclined to identify as sponges almost any irregularly shaped, problematical fossil

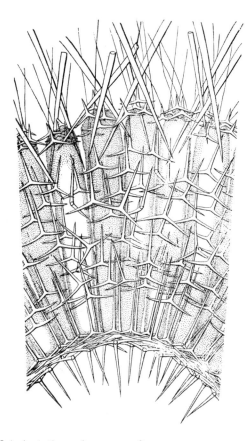

**Above** Spicules in the modern sponge *Sycon*.

collected from marine sedimentary rocks. While it is true that many fossil sponges do have a somewhat amorphous appearance, a great many exhibit very distinctive shapes. For example, the Cretaceous sponge *Tulipa* as the name suggests resembles a tulip on a stalk. At the apex of the tulip is a hole, the osculum, through which filtered water would have been expelled when the sponge was alive. Another Cretaceous genus, *Guettardiscyphia*, has a star-like shape due to its five or six radiating flanges. The peculiar Ordovician sponge *Brachiospongia* is equipped with a series of finger-like props that apparently supported it on the sea bed. Fossil sponges shaped as spheres can also be found. They include *Porosphaera*, which is abundant in the Cretaceous Chalk of England. Many examples of *Porosphaera* have a single hole passing all the way through them. Bronze Age inhabitants of the UK took

advantage of this feature and threaded together *Porosphaera* fossils to make necklaces.

Some sponges secrete acids enabling them to excavate a series of linked chambers in shells or limestone. These boring sponges are extremely important agents of bioerosion, destroying the shells of living molluscs and corals, and eating into limestone rocks exposed along tropical coastlines. Fossil borings are known by the trace fossil name *Entobia* and first appeared in the Jurassic but did not become abundant until the late Cretaceous.

As mentioned above, a basal skeleton of calcium carbonate is secreted in some groups of sponges. Best known among these are the stromatoporoids. These layered, mound-like or branching sponges are particularly common in shallow water deposits of Silurian and Devonian age. Individual stromatoporoids could grow to over 1 metre in height, as on the Swedish island of Gotland where eroded Silurian stromatoporoids form small sea stacks on the foreshore. Spicules are absent, or at least have yet to be found, in Palaeozoic stromatoporoids and, until recently, there was serious doubt about the biological affinities of the group; most scientists believed that stromatoporoids were cnidarians related to modern hydrozoans. However, the existence of radiating channels on the surfaces of some stromatoporoids (called astrorhizae) supports their identity as sponges. Astrorhizae have no analogues among Recent cnidarians but very similar structures are found in the Recent sponge *Ceratoporella*. By comparison with *Ceratoporella*, astrorhizae can be inferred as the sites of oscula where filtered water was vented away from the surface of the animal. Another group now regarded as sponges with basal calcareous skeletons are the chaetetids which were formerly assigned to the tabulate corals. However, spicules diagnostic of a sponge affinity have been found embedded in the skeletons of a few species of chaetetids.

Both stromatoporoids and chaetetids were important reef builders at various times during the Palaeozoic.

## SPONGE CLASSIFICATION

There are three classes of sponges living today: Hexactinellida (glass sponges), Demospongiae and Calcarea (calcisponges). It is possible to place all of the extinct groups of sponges into one of these three extant classes, with the exception of the Archaeocyathida which may warrant recognition as a separate class. Unfortunately, the superficial appearance of fossil sponges is not usually a good indicator of which class they belong to; similar shaped sponges can occur in all three classes. Spicules are the best indicator of class membership but are not always easily visible in fossil sponges. Hexactinellids have siliceous spicules with six rays, demosponges possess siliceous or collagenous spicules with diverse shapes, and calcisponges secrete spicules of calcite. Sponges with calcareous basal skeletons, which may consist of either calcite or aragonite, were once placed into a fourth class (Sclerospongia) but are now considered by many specialists to be more appropriately distributed among the demosponges and calcisponges. Stromatoporoids and chaetetids, the two geologically important groups, are both demosponges. However, it is common for the stromatoporoids to be placed in their own class, Stromatoporoidea.

Ancient reefs were also often constructed by sponges lacking calcareous basal skeletons. Among the largest is the Permian Capitan Reef in Texas with a reef front some 500 km (300 miles) long standing 600 m (2000 feet) above the basin. At the opposite end of the size spectrum are small reefs constructed by the sponge *Platychonia* in the mid Jurassic cliffs of Normandy in France, the same cliffs containing the screw-like bryozoan *Terebellaria*. These reefs had little elevation above the level of the surrounding sea bed but none the less formed havens for a diverse community of brachiopods, bivalves, bryozoans, worms and other sponges. Although *Platychonia* secreted siliceous spicules, these spicules were dissolved away before or soon after burial of the sponge reef.

Dissolution of siliceous spicules is more common than might at first be expected in view of the seemingly durable nature of silica. Dissolved sponge spicules are the main source of silica forming the concretionary flints that occur in the chalk deposited over an enormous part of Europe and Asia during the late Cretaceous. Flints were formed by the re-precipitation of this silica from solutions passing through the fine-grained sediment. In some cases a sponge is found in the core of the flint when it is cracked open, but most flints contain no obvious remains of the sponges from which they were derived. Flints do however commonly encase other fossils such as echinoids and molluscs, often because these fossils formed nuclei around which the silica was originally precipitated.

**Above** A Cretaceous flint broken open to reveal the sponge *Ventriculites* with a root-like base.

Among the oldest fossil sponges are the archaeocyaths that flourished briefly during the Cambrian. For a long time the relationship of archaeocyaths to modern animals was unknown but, despite lacking spicules, it is now believed that they are an extinct group of sponges. Most archaeocyaths are shaped like an inverted cone, open at the wider, top end. The cavity of the cone is thought to be a spongocoel, and the double layer of walls enclosing it are porous, compatible with the filtering function of a sponge in which water passes through the body. Archaeocyaths are important because, together with other organisms such as algae, they were capable of constructing small reefs, the oldest known reefs involving multicellular animals. Reef-building archaeocyaths are mostly modular (colonial) species. They initiated a pattern, which has continued through to the present day, for colonial animals to be the main organisms building organic reefs in the sea.

### ACTINOSTROMA: DEVONIAN, WORLDWIDE

Generally dome-shaped, this stromatoporoid sponge lacks spicules but has a robust basal skeleton made of calcite with a compact microstructure. In vertical section the skeleton is seen to comprise pillars, oriented perpendicularly to the sponge surface, linked by horizontal elements called colliculi. Concentric growth

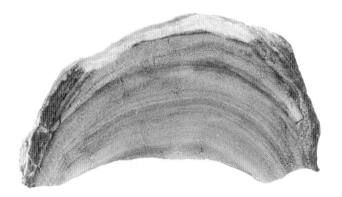

**Above** Vertical polished section through the dome-shaped stromatoporoid *Actinostroma*, 9 cm in width, from Devonian rocks of Western Australia.

lines may be visible on the epithecal wall forming the basal underside of the sponge.

*Actinostroma* is fairly representative of the stromatoporoids, an exceedingly important group of sponges in the fossil record, especially the Silurian and Devonian where they are significant reef-builders.

### ASTRAEOSPONGIUM: ORDOVICIAN–DEVONIAN OF EUROPE AND NORTH AMERICA

**Above** Stout six-rayed spicules are visible covering the surface of this 6.5 cm wide specimen of the sponge *Astraeospongium* from the Silurian of Tennessee.

This sponge is generally saucer shaped and 3–5 cm (1–2 inches) in diameter. The skeleton lacks canals, and is thick walled, without any trace of an attachment structure. On the surface is a mass of large spicules, each generally having six long rays in a plane parallel to the surface of the sponge, producing star-like patterns, plus two shorter perpendicular rays. Spicule size increases from the concave upper surface to the convex basal surface.

*Astraeospongium* belongs to the Calcarea, more specifically to a small Palaeozoic order called the Heteractinida. Sponges of this order were apparently unattached, the convex surface of the saucer-like body resting on the sea bed in quiet water environments.

## *BARROISIA*: JURASSIC–CRETACEOUS OF EUROPE

A simple or more often modular sponge (see colour fig. 7) made of finger-like branches a centimetre or less in diameter. Fractured branches show an internal structure comprising a stack of chambers penetrated centrally by a tube that opens at the tips of the branches as an osculum. Externally, branch surfaces are porous and may be annulated, reflecting the internal partitioning into chambers. The microstructure of the skeleton consists of calcite fibres containing embedded spicules.

*Barroisia* belongs to the Calcarea and exhibits a 'sphinctozoan' organisation, with a chambered skeleton. Although also having a sphinctozoan organisation, the Recent sponge genus *Vaceletia*, by contrast, is classified in the Demospongia on the basis of other aspects of its anatomy. This underscores the difficulty of identifying sponges from their shape alone. It is likely that living tissue in *Barroisia* occupied only the top few chambers, and water that entered the sponge through the small pores in the sides of the branch walls was filtered for food particles, and vented through the oscula at the branch tips.

## *BRACHIOSPONGIA*: ORDOVICIAN OF NORTH AMERICA

This sponge is large (up to 35 cm (14 inches) in diameter), unrooted, and has 6–12 hollow projections radiating outwards from a tubular or cup-shaped central region with a chimney-like opening. The hollow projections often curve downwards, resembling bent fingers. Small pores may penetrate the smooth or textured external surface, and the thick skeleton is made up of a felted mass of spicules.

The downward-turned projections of the peculiar hexactinellid sponge *Brachiospongia* apparently acted as props supporting the sponge on hardened sea beds.

**Above** Measuring 17 cm in width, this specimen of *Brachiospongia* comes from the Ordovician of Kentucky. The central chimney has broken-off.

## *CHAETETES*: SILURIAN–CARBONIFEROUS OF EUROPE, NORTH AMERICA AND ASIA

*Chaetetes* is typically a large sponge that may be laminar, hemispherical or globular in shape. The skeleton consists of long, narrow tubes (calices), generally 1 mm

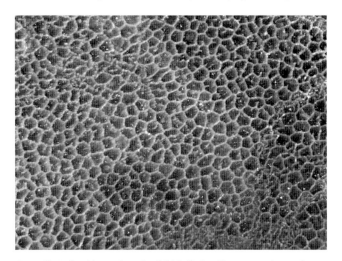

**Above** Part of a thin section of a British Carboniferous specimen of *Chaetetes* showing cross sections of the calices, each less than a millimetre in diameter.

(0.04 inches) or less in diameter, which in cross-section are polygonal or have a more meandering shape. Calices are divided horizontally by partitions (tabulae) and may have spines projecting inwards from their walls. The calcareous walls have a fibrous microstructure, lack pores, and very occasionally contain simple embedded spicules.

Long regarded as a tabulate coral, this genus is now recognised as a sponge with a calcified basal skeleton and is classified within the Demospongia. *Chaetetes* is particularly common in shallow water limestones of the Carboniferous where it sometimes constructs small reefs.

**Above** Several individuals, each a couple of centimetres in diameter, of the sponge *Choia* adorn a bedding plane of Cambrian shale from Utah.

## *CHOIA*: CAMBRIAN OF NORTH AMERICA

Specimens of this small sponge are usually preserved flattened on bedding planes. In life, the animal apparently consisted of a thin, low, cone-shaped disc covered by radiating spine-like spicules, some extending beyond the edge of the disc.

*Choia* is one of several sponge genera that occur in the Burgess Shale of Canada and is also found elsewhere in Cambrian rocks. It is a demosponge interpreted to

have rested on the sea bed with the more convex side uppermost. The covering of spicules would doubtless have discouraged predators.

## *HYDNOCERAS*: DEVONIAN–CARBONIFEROUS OF EUROPE, NORTH AMERICA AND ASIA

This hexactinellid sponge has the form of a tall cone or a vase that is generally octagonal in cross-section and may have a basal root tuft. Marking the angles between the eight sides are ridges with periodic bulbous swellings. Specimens are often found preserved as casts formed by sediment infilling the central cavity (spongocoel). The surface of the cast is impressed with a meshwork of spicules.

*Hydnoceras* inhabited soft bottoms, anchoring itself into the silty sediment using the root tufts.

**Above** Over 17 cm tall, this natural cast of the hexactinellid sponge *Hydnoceras* comes from Devonian rocks of New York State.

## *METALDETES*: CAMBRIAN, WORLDWIDE

**Above** A small block of limestone, 5 cm across, from the Cambrian of South Australia, containing sectioned specimens of the archaeocyathan sponge *Metaldetes*.

A cup-shaped archaeocyathan sponge, *Metaldetes* is usually solitary but sometimes modular with multiple cups. Each cup has two walls separated by a space called the intervallum that is crossed by porous partitions in various orientations. The inner wall surrounds the central cavity of the sponge and is regularly porous. The outer wall consists of an irregularly porous layer covered by a lamina containing very small pores. Spicules are lacking.

*Metaldetes* is one of about 250 genera of archaeocyathans, a distinctive group of sponges that flourished during the Cambrian but did not survive into the Ordovician. Specimens usually occur in limestones where the sponge skeleton may be silicified, making it stand proud of the matrix after weathering.

## *POROSPHAERA*: CRETACEOUS–PALEOCENE OF EUROPE

A small to medium-sized calcarean sponge, *Porosphaera* varies from spherical to pear-shaped, cone-shaped, limpet-shaped or sheet-like when encrusting hard surfaces such as echinoid tests. Canals open on the sponge surface as circular pores 0.15–0.30 mm in diameter. The walls of these canals show a spicular structure close to the sponge surface but this is lost at deeper levels where the spicules become intergrown. Shallow grooves are sometimes visible incising the colony surface.

*Porosphaera* is a very common fossil in the late Cretaceous Chalk of Europe. One particular species, *Porosphaera globularis*, is spherical and frequently has a natural hole passing through the centre of the sponge, probably resulting from growth around the stem of an unpreserved animal or plant. These bead-like fossils were collected by people living during the Bronze Age and strung together as necklaces, among the oldest instances of fossil collecting by humans.

**Above** Spherical specimen of *Porosphaera*, from the Cretaceous Chalk of Wiltshire, England. This 1.7 cm diameter example shows clearly the hole often present in this sponge.

## *RAPHIDONEMA*: CRETACEOUS–EOCENE OF EUROPE AND ASIA

**Above** Oblique view of a 7 cm wide specimen of *Raphidonema*, from the Lower Cretaceous Faringdon Sponge Gravel of southern England.

*Raphidonema* is a cup-shaped or funnel-shaped sponge with thick walls, often knobbly and/or convoluted. Some examples grew cemented to shells or pebbles but others regenerated from broken fragments. The interior surface contains small, and sometimes large, pores. The skeleton is made of spicules fused together into strands or fibres.

An ancient tidal channel in Oxfordshire is filled by up to 50 m (165 feet) of early Cretaceous sediments. This deposit is called the Faringdon Sponge Gravel, reflecting the abundance of sponges of which *Raphidonema* is one of the most common.

## *SIPHONIA*: JURASSIC–CRETACEOUS OF EUROPE

The pear-shaped or spherical main body of *Siphonia* is supported on top of a long, narrow stem with a system of roots at the base. Specimens preserving body and stem may have the appearance of a tulip; indeed one species is named *Siphonia tulipa*. An opening at the top of the sponge is the osculum that leads into the spongocoel, a cylindrical or funnel-shaped depression. Sectioned specimens reveal the internal structure comprising narrow incurrent canals perpendicular to the outer surface of the sponge that link with broader excurrent canals arranged parallel to the surface of the sponge and emptying into the spongocoel. The large four-rayed spicules are interlocked to form a rigid reticulate skeleton.

Found especially in the Cretaceous Chalk, this demosponge is frequently replaced by flint.

**Above** Encrusted by oysters, this 4.5 cm wide example of the sponge *Siphonia* was collected from Cretaceous rocks in Devon, England.

# GRAPTOLITES

An age-old student prank is to try to pass off a pencil line drawn on a piece of slate as a fossil graptolite. This deception is possible because of the carbonaceous preservation and pencil-line appearance typical of many graptolites found in Ordovician and Silurian shales. Indeed, these extinct invertebrates have been dubbed 'the writing in the rocks', and the name graptolite is derived from the Greek for 'written animal'. To extend the writing analogy further, graptolites are of great geological utility because they spell out the age of their host rocks with remarkable precision. Graptolites rank with ammonites as being the best biostratigraphical macrofossils of all, some species defining zones of less than 1 million years in duration.

But what exactly are graptolites? For a long time the affinity of these Cambrian–Carboniferous colonial animals was unknown. They were commonly believed to be either cnidarians or bryozoans, until comparative studies demonstrated their close relationship with an obscure group of living marine animals called the pterobranch hemichordates. Only two genera of pterobranchs exist today, *Rhabdopleura* and *Cephalodiscus*. Of these, *Rhabdopleura*, which has a fossil history stretching back to the Cambrian, has provided palaeontologists with crucial evidence for understanding the biology of the extinct graptolites. The branching colonies of *Rhabdopleura* consist of small zooids housed in long tubes. Each zooid has a lophophore with two arms bearing a series of tentacles, a mouth leading into a U-shaped gut, and a muscular pad called the cephalic shield. As in the unrelated bryozoans, extension of the zooid from its tube is accomplished by hydrostatic pressure and withdrawal into the tube by contraction of muscles. However, zooids of *Rhabdopleura* are also able to crawl within their tubes using the muscular cephalic shield. New zooids are budded at the tips of the colony branches from a stolon that forms a link between all of the mature zooids. By contrast, in the other living pterobranch, *Cephalodiscus*, the mature zooids sever their connection with the stolon and are able to wander over the colony surface freely. They can even vacate their tube, migrate to another part of the colony and make a new tube in this location. Both genera of extant pterobranchs are suspension feeders and subsist on plankton. They use cilia to draw water towards the lophophore, trap food particles in a mucus net stretched between the tentacles, and transport the captured particles along grooves in the arms and on to the mouth.

The Class Graptolithina consists of several orders. Of these the most significant are the Dendroidea and Graptoloidea, the dendroid and graptoloid graptolites. Dendroids appeared in the mid Cambrian and persisted in relatively small numbers into the late Carboniferous. They have bushy colonies with many branches (stipes) and zooids exhibiting polymorphism. Compared to the evolutionarily conservative dendroids, graptoloids rode a rollercoaster of evolutionary innovation, radiation and extinction between their first appearance in the early Ordovician and final disappearance in the early Devonian. Graptoloids have more regularly shaped colonies than dendroids with fewer stipes (often only one) and limited or no zooidal polymorphism. It is the graptoloids that have attracted most interest from palaeontologists, not only because of their greater abundance but also because of their utility in stratigraphy and the challenge of understanding their unusual ecology. Dendroids typically lived as sessile suspension-feeders rooted to the sea bed, an ecology similar to that of modern *Rhabdopleura* as well as some reticulate bryozoans that have a very similar colony form. Graptoloids evolved from dendroids and inherited a free-living planktonic ecology from their immediate dendroid ancestor, thought to be the genus *Rhabdinopora*. They are thus unlike any of the other main colonial groups represented in the fossil record. No close modern analogues of graptoloids exist. However, pelagic colonies do occur in other phyla. A group of planktonic sea-squirts (Phylum Urochordata) called salps can form colonial chains that move by jet propulsion, and colonies of siphonophores (Phylum Cnidaria) may employ fish-like, serpentine movements to swim.

The colonial skeleton, often referred to as a rhabdosome in both dendroids and graptoloids, originates from the sicula, the conical skeleton, commonly 1–2 mm (0.04 inches) long, of the first zooid in the colony. The tubular skeletons of later budded zooids are called thecae and are often arranged in one or two series, occasionally four series, along the branches (stipes) of the colony. Thecae are typically spaced about 1 mm (0.04 inches) apart along the stipes, although this distance varies in different species. A hollow rod called the nema forms the 'backbone' of the stipes.

In dendroid graptolites the stipes may be linked by transverse dissepiments or by the coalescence of adjacent stipes. Polymorphism of thecae occurs in dendroids but is usually lacking in graptoloids. The smaller thecae (bithecae) of dendroids have been interpreted as male zooids and the larger thecae (authothecae) as female zooids. It is often assumed that the evolutionary loss of this polymorphism in graptoloids signifies a change from zooids of separate sex to hermaphroditic zooids. A substantial increase in the size of the thecae occurs from the oldest (proximal) to youngest (distal) parts of graptoloid colonies. This reflects the budding of progressively larger zooids as the colony grew, a characteristic seen in many colonial animals.

The stipes of graptoloid colonies may be single, two in number, or more numerous. Some stipes are straight, some gently curved, and others spiral in shape. The well-known graptoloid *Didymograptus murchisoni* from the mid Ordovician has two stipes arranged like a tuning fork. So-called multiramous graptolites with numerous stipes may add stipes by bifurcation (e.g. *Anisograptus*), or by the development of new lateral stipes called cladia at the edges of existing stipes. A particularly distinctive Silurian genus is *Cyrtograptus* with a spiral primary stipe producing cladia at intervals along its outer edge. In all multiramous graptoloids, the symmetry of the colony was maintained as the colony grew by stipe elongation

and multiplication. This was necessary to preserve the balance of the colony in the plankton. Graptoloids having a single stipe include uniserial genera (e.g. *Monograptus*) with thecae opening along one side of the stipe, biserial genera (e.g. *Diplograptus*) with thecae on the two opposite sides of the stipe, and quadriserial genera (e.g. *Phyllograptus*) with four columns of thecae oriented at 90 degrees to each other.

An evolutionary trend towards reduction in the number of stipes has long been recognised in graptoloids. The dominantly multiramous graptoloids of the early Ordovician were succeeded by species with two stipes. Silurian graptolite faunas consist overwhelmingly of species of monograptids with a single stipe. Monograptids evolved during the latest Ordovician and survived the mass extinction at the end of that period to radiate explosively in the Silurian. The reason for this evolutionary trend is unknown although it may relate, at least in part, to the elimination of interference between zooids on adjacent stipes that often come into close contact in larger multiramous colonies. It may also be that graptoloids with a single stipe were better adapted to turbulent water.

Most graptoloids are between 1 and 10 cm (0.5–4 inches) in length. There are, however, some gigantic graptoloids of which the largest reported is an incomplete colony 145 cm (57 inches) long found in the Silurian of Wales. This colony would have contained about 1200 zooids and may have reached at least 25 years in age, based on a similar growth rate to living *Rhabdopleura* colonies. In contrast, *Corynoides* is minute, consisting only of a sicula and three or four thecae.

The morphology of graptoloid thecae – the skeletons of the individual zooids – is used extensively in taxonomy. At their simplest, thecae are straight-sided tubes inclined at an angle to the stipe so as to give a saw-tooth appearance. More complex thecae may have hooked ends, be provided with lappets that greatly constrict the terminal aperture, or have terminal spines. All of these morphologies can be tentatively interpreted as defences against predators. Although scarcely anything is known about predation on graptoloids, by analogy with extant colonial animals such as bryozoans, it is likely that they had some predators whose strategy was to attack one zooid at a time via the aperture. This would be rendered much more difficult in graptoloids with hooked thecae, constricted apertures or spines surrounding the aperture.

As mentioned above, graptolites are often preserved as carbonaceous impressions. They can also be preserved as internal moulds (steinkerns) by pyrite infilling the

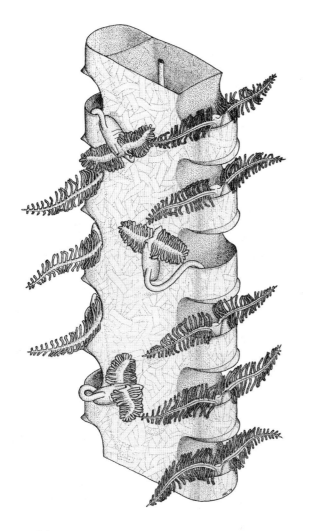

**Above** Life reconstruction of a biserial graptolite colony. Three of the zooids are depicted plastering bandages of scleroprotein onto the outside of the colony, the others are in their normal feeding positions.

thecae, or as pale films of phyllosilicate minerals. The most revealing specimens, however, are preserved in three-dimensions in cherts and, more particularly, in limestones from which they can be released by acid treatment. It is material of this sort that has been invaluable in elucidating the fine structure of the graptolite skeleton. The original skeleton of graptolites was not mineralised but instead was composed of an organic scleroprotein. Scanning electron microscopy of exceptionally preserved graptolites has revealed the bandaged appearance of the outermost skeleton. Strips of scleroprotein were evidently plastered onto the colony by zooids stretching out from their thecal homes. It is thought that the bandages were secreted from the cephalic discs of the zooids, the same organ that is used by zooids of extant *Rhabdopleura* to permit crawling movements over the colony surface. Beneath this bandaged layer the earlier-formed part of the thecal skeleton comprises growth increments, called fusellae, consisting of rings or half-rings of skeleton that join together at zig-zag sutures. Very similar structures occur in *Rhabdopleura*. Extreme reduction of the skeleton occurs in a peculiar group of graptoloids called retiolitids. The thecae of retiolitids are represented by a network of narrow struts and girders.

There has been considerable debate about the ecology of graptoloids, centred on the question of whether they propelled themselves in the plankton or were passive drifters. Palaeontologists advocating 'automobility' have suggested that the ciliary feeding currents of the zooidal lophophores could have acted in unison and were of sufficient strength to enable the colony to move through the water. Analogies with planktonic animals living today provide some support for this idea. Modern planktonic animals commonly undertake diurnal migrations, often taking advantage of gravity to descend passively and using active swimming to ascend. At the present day, pelagic colonies of both sea-squirts and salps are able to swim. As noted above, these may be the closest living ecological analogues available for graptoloids. Opponents of the automobility hypothesis have pointed to the likely small size of the zooidal lophophores and feeble currents (based on comparisons with extant pterobranchs) relative to the bulk of the colony. The fact that the varying orientations of thecae in some species would have resulted in zooids pulling in different directions, and the lack of confinement of the zooids within their thecae as evidenced by cortical bandaging, also seem to disfavour automobility. However, a decisive test of the two options for graptoloid ecology is awaited and it may be that some species were automobile but others passive drifters.

In several graptoloid species, clusters of colonies are occasionally found joined together at their proximal ends to give a compound structure called a synrhabdosome. The significance of synrhabdosomes is unknown. It has been speculated that they represent: (1) temporary aggregations of adult colonies for the purpose of sexual reproduction, or to increase food-gathering ability at times of low plankton levels; (2) the result of asexual cloning of a colony; and (3) the association of several colonies sharing a common floatation device or substratum. Automobility in this case is difficult to envisage in view of the radial arrangement of the colonies making up the synrhabdosome.

Despite their planktonic lifestyles with the potential for global dispersal around the oceans of the world, species of graptoloids do show provincialism, with distinct species characterising particular geographical regions. The degree of provincialism varied through time. In the early Ordovician graptoloids were relatively cosmopolitan but during the mid Ordovician they were spilt between two main provinces, Atlantic and Pacific. A more cosmopolitan pattern of distribution was restored in the late Ordovician but provincialism reappeared in the early Silurian. Several factors could have been responsible for graptoloid provincialism in the mid Ordovician, including sea temperature, salinity and current flow. In today's oceans such factors define different water-masses characterised by unique suites of planktonic species. Palaeogeographical reconstructions show that the Pacific graptoloid faunal province lay almost entirely within the ancient tropics, between 30°N and 30°S, whereas the Atlantic province mainly occupied cooler waters outside the tropics. The mass extinction at the end of the Ordovician was due, at least in part, to global cooling; good evidence exists for glaciation at this time. Graptolites in the tropical Pacific province were particularly hard hit by this mass extinction, as would be

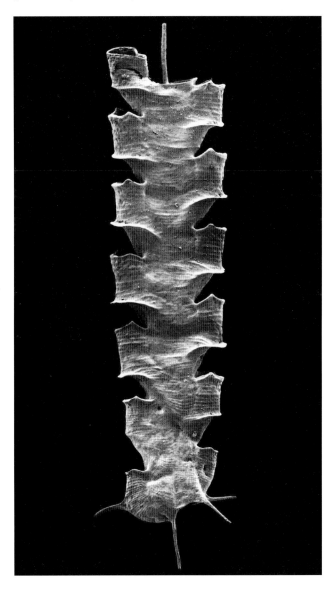

**Above** Taken with a scanning electron microscope, this image of the Ordovician graptolite *Amplexograptus*, about 4.5 mm long, shows the thecae and bandaged skeleton.

expected if areas of warm oceanic waters contracted.

Extinction of the last planktonic graptoloid species occurred in the early Devonian. This did not coincide with a major mass extinction event as it considerably predated the mass extinction of the late Devonian that decimated many other groups of marine animals. Various explanations for graptoloid extinction have been proposed. Floating graptoloid colonies may have been easy targets for predatory fish that were diversifying in the sea at this time. Although difficult to test, an intriguing suggestion is that graptoloids did not really suffer extinction in the early Devonian but instead stopped secreting a skeleton and therefore vanished from the fossil record. If this idea is correct, graptolites may have survived for an unknown length of time after the youngest known fossil examples.

## CALLOGRAPTUS: CAMBRIAN–CARBONIFEROUS, WORLDWIDE

*Callograptus* has irregular colonies that are conical or fan-like in overall shape and up to 15 cm (6 inches) in size. The wide branches bifurcate periodically and in some species also anastomose or are linked by dissepiments. Individual thecae can be difficult to discern, especially in thickened basal parts of the colony where a holdfast may be present.

This long-ranging genus is a dendroid graptolite that had a benthic ecology, living attached to objects on the sea bed using a basal holdfast.

## CLONOGRAPTUS: ORDOVICIAN, WORLDWIDE

This genus has a distinctive flat, multibranched colony, beginning from two primary branches growing in opposite directions and followed by up to nine successive orders of branch bifurcation. The distance between bifurcations increases during growth, and there is a tendency for daughter branches to bend towards one another. Autothecae are denticulate or isolated, and bithecae are present in some species of the genus.

**Above** Colony of the dendroid graptolite *Callograptus* in a block of shale, 6 cm wide, from the Carboniferous of Pendle Hill, Lancashire, England.

**Above** Slightly less than 3 cm wide, the graptolite *Clonograptus* has a characteristic branching pattern, as seen in this specimen from the Ordovician of Newfoundland.

The pattern of branching, with ever greater distances between bifurcations and bending of daughter branches towards one another, results in the efficient filling of two-dimensional space without branches colliding.

**Above** Colony of the graptolite *Cyrtograptus* from the Silurian of Bohemia. Field of view 5 cm wide.

## *CYRTOGRAPTUS*: SILURIAN, WORLDWIDE

The monograptid *Cyrtograptus* has a spirally-coiled main branch with thecae opening on the convex outer side. Secondary branches, called cladia, also arise from the outer side of the main branch and may themselves produce new branches. In all, the colony can be as much as 60 cm (23 inches) in maximum width. The thecae have spines at their apertures.

Experimental studies using models of this graptolite suggest that the floating colonies cork-screwed in the sea, thereby making an efficient unit for sweeping planktonic food from a vertical column of water.

## *DICRANOGRAPTUS*: ORDOVICIAN, WORLDWIDE

This Y-shaped graptolite is usually no more than 4 cm (1.5 inches) long but sometimes grows to over 10 cm (4 inches) in length. The early-formed, proximal part of the colony is biserial, with two rows of thecae back-to-back. These rows separate in later parts of the colony to form two diverging uniserial branches in which the thecae may be spirally arranged around the branch. The thecae vary in shape between species, overlapping or isolated, curved and often having spines.

A dramatic change in colony construction occurs during growth of *Dicranograptus* with the apparent 'unzipping' of a single biserial branch to form two uniserial branches. This presumably signifies an abrupt change in mode of life, for example, in the depth or orientation of the colonies in the water column.

**Above** A typical Y-shaped colony of *Dicranograptus* from the Ordovician of Scotland, measuring about 8.5 cm long.

**Above** Block of siltstone from the British Silurian containing numerous examples of *Monograptus*, each 1.5-2 cm long.

## *DIDYMOGRAPTUS*: ORDOVICIAN, WORLDWIDE

Generally 2–8 cm (1–3 inches) long, colonies of this genus have two branches (stipes) that are variously arranged relative to one another. In many species, the two branches are pendent, forming a tuning fork-like colony with the thecae on each branch facing inwards (see colour fig. 8). In other species they grow away from each other (extensiform). The thecae are simple and tubular, slightly curved or straight.

A large number of twin-stiped species have been referred to this genus but it is likely that they are not all closely related. *Didymograptus* is perhaps the classic example of a planktonic Ordovician graptolite, occurring in great abundance in some shales.

## *MONOGRAPTUS*: SILURIAN–DEVONIAN, WORLDWIDE

Colonies of *Monograptus* comprise a single branch with thecae running along one side only (uniserial). The branch is straight, gently curved or even helicospirally twisted, and can be very long, in exceptional instances over 1 metre. The thecae may be straight or hooked, and

are sometimes spinose. Their size usually increases along the branch, matched by a gradual increase in branch width, and their shape may also change.

This genus has proved to be very useful in the correlation of Silurian rocks. Different species of *Monograptus* are zonal indicators for parts of the Silurian, lending their names to particular zones, e.g. the *Monograptus atavus*, *Monograptus crispus* and *Monograptus tumescens* zones.

## *PHYLLOGRAPTUS*: ORDOVICIAN, WORLDWIDE

The leaf-like colonies of this robust graptolite consist of four branches joined back-to-back along their entire lengths. Each branch is deep and oriented at right angles to its neighbours. Thecae are straight or slightly curved, strongly overlapping with adjacent thecae in the same branch.

The presence of a virgella (a spine-like rod projecting from the first-formed individual) in *Phyllograptus* points to a planktonic mode of life. This accords with the wide geographical distribution of the genus that makes it useful as a zonal fossil for global correlation of Ordovician rocks.

**Above** *Phyllograptus* crowded onto a bedding plane. The field of view in this specimen from the Ordovician of Quebec is 6 cm wide.

**Above** The thecae are almost 1 cm long in this fragment of the graptolite *Rastrites* from the Bohemian Silurian.

## *RASTRITES*: SILURIAN, WORLDWIDE

This monograptid is distinguished by the extremely long thecae. Colonies are small, typically less than 4 cm (1.5 inches) long, and the single branch is usually curved or hooked. Thecae arise from the branch at about 90°, are widely spaced and may be up to 2 cm (1 inch) long in some species. They often end in hooked or hooded apertures.

Emerging from the ends of the very long thecae, the zooids must have been more spatially isolated from one another in this genus than was typical for graptolites, although they were presumably still linked internally by the stolon.

## *RHABDINOPORA*: ORDOVICIAN, WORLDWIDE

The reticulate colony of *Rhabdinopora* is conical and up to 10 cm (4 inches) long. Branches (stipes) are parallel, bifurcate at intervals, occasionally anastomose, and are cross-linked with adjacent branches by dissepiments. The founding theca (sicula) has a spine-like nema. Two types of theca are present: autothecae, commonly with spines, and smaller bithecae.

At first sight *Rhabdinopora* bears a remarkably close resemblance to the bryozoan *'Fenestella'*, both having

**Above** Reticulate colony of the Cambrian graptolite *Rhabdinopora*, 5.5 cm high, collected in Herefordshire, England.

reticulate colonies composed of branches linked at intervals by dissepiments. However, the composition of the skeleton is quite different, organic in *Rhabdinopora* but calcareous in *'Fenestella'*. There are also many detailed differences at zooidal level. In addition, whereas *'Fenestella'* is a benthic animal, firmly attached throughout its life to a substrate on the sea bed, *Rhabdinopora* is a primitive example of a planktonic graptolite that floated freely in the ocean.

# 3

# Shells galore

SHELLS OF MARINE ANIMALS represent the most common invertebrate fossils. Most fossil shells belong to one of two major groups or phyla, the Mollusca (molluscs) and the Brachiopoda (brachiopods). Both molluscs and brachiopods have existed in great abundance in the seas of the world for more than 500 million years. During this time they have produced huge numbers of resistant and geologically stable shells made generally of calcium carbonate and well suited for fossilisation. Sometimes these shells are widely scattered through the sedimentary rocks in which they occur. On other occasions they are packed densely into discrete horizontal bands called shell beds, or form shelly limestones that can attain considerable thicknesses. Fossil shells from poorly lithified rocks can often be extracted in pristine condition and may be virtually indistinguishable from the shells of modern animals. The fossil examples are, however, usually slightly heavier due to the introduction from solution of diagenetic minerals into tiny spaces within the fabric of the shell. In addition, fractured surfaces of fossil shells tend to have a white, chalky texture that contrasts with the transluscent appearance of modern shells.

**Left** Shell of the Eocene gastropod *Crisposcala* from France, about 2 cm high.

Furthermore, any coloration present in the shell is normally lost during fossilisation, although fossil shells do occasionally preserve vestiges of their original colours. With a few exceptions, the shells of molluscs and brachiopods lie externally to the soft body; the shells function primarily to protect the vulnerable soft parts of the animals situated within. In some molluscs the shells are single or univalved, as in gastropods and cephalopods, but many molluscs and all brachiopods have shells comprising two distinct valves, the bivalved condition. A minor group of molluscs called chitons or polyplacophorans have shells consisting of eight articulated plates. Growth of the shell in all of these animals occurs by accretion, new shell material being added progressively to the outer edge of the valves. A crude analogy can be made between shell growth and the way in which a knitted sock grows in length as the knitter stitches more and more wool around the open end. Also like a sock, each valve of the shell normally has a closed end and an open end, corresponding respectively with the earliest and latest formed parts of the shell.

Geometrically, the valves of molluscs and brachiopods are all fundamentally cone-shaped (with the exception of the polyplacophoran molluscs). The cones differ from species to species mainly in the abruptness with which the cone expands and the extent to which it is twisted and curved. An elegant pioneering application of computer simulation of organic growth was undertaken by the American palaeontologist David Raup. Raup was interested in the extent to which shells in nature adopted all of the theoretically possible shapes. He was able to replicate shell shape using a simple program that started with a cone and varied four geometrical parameters: the rate at which the cone expanded in width as it grew in length, the tightness of spiral coiling, the 'translation' of this coiling into a third, perpendicular axis to give a helicospiral (screw-like shape), and the distance of the whorl from the axis of coiling. In true shells, slight differences in the amount and direction of accretion of new shell material around the open edge of the cone determine how these four parameters vary. These differences are sufficient to bring about major changes in shell shape; for example, from the cap-like limpet with its extreme whorl expansion to the almost straight and narrow tusk shell with its limited whorl expansion, or from the near one-dimensional coiling of a ramshorn snail shell to the highly three-dimensional coiling of a whelk shell.

## MOLLUSCS

With a history extending back to the Cambrian, molluscs are extremely common in the fossil record, becoming ever more abundant in younger rocks. Their diversity in the Cenozoic is unsurpassed by any other macrofossil group. There are an estimated 50,000 species of molluscs around today, and it is thought that as many as 60,000 fossil species have been described. The overwhelming majority of molluscs have mineralised shells, hence their superb fossil record. These immensely successful invertebrates have colonised every major environment except the air. Molluscs can be found living at all depths in the sea, from intertidal to abyss, and also inhabit freshwater rivers and lakes and the land surface. An important food item for humans, molluscs such as clams, oysters, scallops, whelks and squid grace the menus of seafood restaurants all around the world.

## MOLLUSC CLASSES WITH GEOLOGICAL RANGES

Class Caudofoveata (Recent)
Class Solenogastres (Silurian–Recent)
Class Polyplacophora (Cambrian–Recent)
Class Monoplacophora (Cambrian–Recent)
Class Bellerophonta (Cambrian–Triassic)
Class Gastropoda (Cambrian–Recent)
Class Cephalopoda (Cambrian–Recent)
Class Scaphopoda (Carboniferous–Recent)
Class Bivalvia (Cambrian–Recent)
Class Rostroconchia (Cambrian–Permian)

There are three major and several minor classes of molluscs. Bivalves (clams and their relatives), gastropods (snails and slugs) and cephalopods (octopus, squid, nautilus, etc.) are the three dominant molluscan classes, not only today but also in the fossil record. The most primitive molluscs, however, belong to two obscure classes (Caudofoveata and Solanogastres; together sometimes called Aplacophora) which have worm-like bodies, lack a solid shell and are very seldom fossilised. Five other minor classes that can be found as fossils are polyplacophorans (chitons or amphineurans), monoplacophorans, bellerophontids, scaphopods (tusk shells) and rostroconchs. A series of anatomical features unite these major and minor classes and allow them to be distinguished from every other invertebrate phylum. All molluscs have fleshy bodies in which the coelom (fluid-filled cavity) is greatly reduced. An open circulatory system – the haemocoel – is present, as are distinctive gills called ctenidia used for respiration and in some species feeding too. The body is covered by a thick skin or mantle that secretes the shell of calcium carbonate in those species having shells. Many molluscs also possess a rasping, file-like structure called the radula used during feeding, for example by limpets to scrape algae off rocks.

# Bivalves

Molluscs with two distinct valves belong to the Class Bivalvia, sometimes also referred to by the alternative names Pelecypoda and Lamellibranchia. There are over 8000 living species of bivalves, most of which can be distinguished from one another by variations in shell shape. The two valves are joined along a hinge-line. In most species a series of teeth in one valve interlocks with sockets in the other. The soft tissues of the animal are typically enclosed entirely within the space between these two valves, although in some bivalves soft tissues extend permanently beyond the confines of the valves. In the majority of bivalves the left and right valves forming the shell are almost mirror images of one another. However, the individual valves themselves are asymmetrical; a line drawn from the curved umbo, marking the point of commencement of shell growth,

vertically to the outer margin, divides the valve into two parts that are unequal. Internally, one or two adductor muscles run from the surface of one valve to the opposite surface of the other valve. Contraction of these muscles brings the valves together, tightly closing the gape of the shell in the majority of species. This closure protects the soft parts of the animal by making it very difficult for predators in the wild (or humans in a restaurant) to prise apart the two valves. The animal opens its own shell by relaxing the adductor muscles, permitting an elastic ligament situated at the hinge-line to force the valves apart like a spring. The effects of the ligament can be seen in dead bivalve shells that are still articulated. Because muscle deterioration precedes the loss of ligament elasticity, such shells are commonly

**Below** Interior of a 12 cm long shell of the bivalve *Panopea* from the Miocene of Virginia, showing the pallial line with its deep sinus (arrowed) .

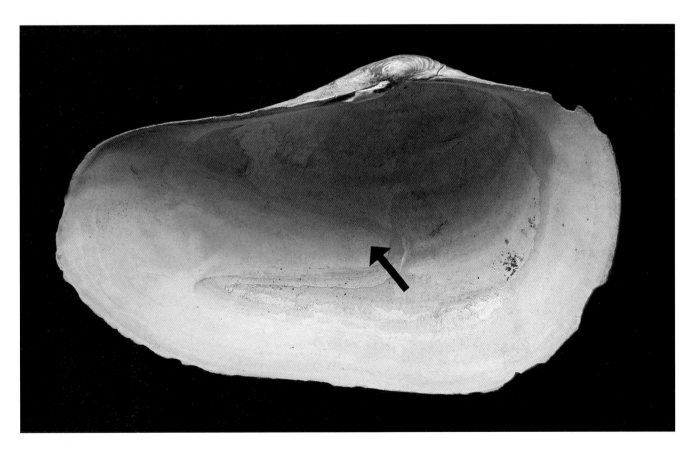

found splayed open, 'butterflied', through the residual elastic force of the ligament.

Four main groups of bivalves exist: protobranchs, anomalodesmatans, pteriomorphs and heterodonts. Conservative in their morphology, the protobranchs are a primitive group of about 600 living species that are particularly abundant in the deep sea. They can be traced back in geological time to the Cambrian. All protobranchs have rather small shells composed of the calcium carbonate mineral aragonite, generally lacking surface ornament. Protobranchs (Superfamily Nuculoidea) are termed 'nut-shells', as they resemble a pistachio nut in shape. Like a great many other bivalves, protobranchs are infaunal, living within the sediment on the sea bed. However, protobranchs are relatively unusual among infaunal bivalves in being predominantly deposit feeders. They feed by processing the sediment in which they live to extract organic material, in much the same way as an earthworm gains

**Above** The two, unequal-sized valves of a post-Pliocene specimen of the spiny bivalve *Spondylus*, the larger valve measuring 5.1 cm in height.

nutrition from the soil. Protobranchs, and other infaunal bivalves, burrow through the sediment using a muscular foot that is protruded beyond the margins of the shell.

Another group of bivalves, the anomalodesmatans, range from the Ordovician to the present day. These too have aragonitic shells, although often studded with spicules on the outer surface, and also burrow into the sediment. Unlike protobranchs, anomalodesmatans burrow not for the purpose of obtaining food but rather for protection. Anomalodesmatans are mostly suspension feeders that filter plankton-laden water sucked into a siphon connecting the buried shell to the sediment surface. Some anomalodesmatans are very common in the fossil record. For example, the genera *Pleuromya* and *Pholadomya* abound in shallow water marine sediments of the Jurassic. They are most often found articulated but as steinkerns (sediment cores), totally bereft of the thin aragonitic shell that was dissolved after the animal died. Like their modern relatives *Laternula* and the wonderfully named *Pandora*, both *Pleuromya* and *Pholadomya* were deep burrowers, living well below the surface of the sea bed. Deep burrowing demands the presence of long siphons. Although the siphons are not usually fossilised in bivalves, an indication of their former length can be obtained from the size of a notch, called the pallial sinus, on the inner surface of the shell. The pallial sinus is an inflection of the line along which the soft tissues of the animal were permanently fixed to the shell; a deep pallial sinus normally indicates long siphons and deep burrowing. Elongation of the shell also often points to a deep burrowing lifestyle in bivalves.

In distinct contrast to the burrowing anomalodesmatans discussed so far, genera such as the late Palaeozoic *Permophorus* were epifaunal; they lived on the sediment surface, probably tethered and kept in place by a structure called the byssus. The third bivalve group, which are known as pteriomorphs, include many genera that are also provided with a byssus for anchorage. The common mussel *Mytilus edulis* is probably the most familiar of pteriomorph bivalves. Living clustered in dense aggregations attached to rocks in the intertidal zone, the byssus of mussels comprises numerous threads resembling thin nylon ropes. These threads are immensely strong and can be secreted very rapidly by the animal such that dislodged individuals are able to become reattached quickly. The risk of dislodgement is one of the main hazards of a mussel's life, particularly for those individuals situated on the outsides of the clusters; individuals nearer the centre of the clusters benefit by being protected from the buffeting of waves by those around them. Another cause of mussel death are predatory starfish. These attach themselves to the shell using their tube-feet, gradually pulling the valves apart before moving their stomachs over the soft parts of the mussel and digesting them. Other pteriomorphs have evolved ways of making predation by starfish more difficult. Spondylids have shells covered by spines. Although the spines themselves may act as a deterrent against predators, more importantly they have the effect of encouraging sponges to colonise the shell surface. The covering of sponges makes it very difficult for the tube-feet of the starfish to gain purchase on the shell. Another group of pteriomorphs, the oysters, obtain a degree of protection from predators by being firmly cemented to hard surfaces, such as rocks, hindering their manipulation by predators.

Oysters are also bivalves and are as common as fossils, especially in rocks of Jurassic and younger age. There are several reasons for their great abundance: oyster shells are often very thick and are made of the resilient calcium carbonate mineral calcite; living oysters are capable of inhabiting a range of different environments varying in salinity from brackish to fully marine; and oysters often live in very dense populations, with young shells cementing to the surfaces of older shells. Fossil shell beds composed almost entirely of oysters are not uncommon, some representing reef-like submarine banks on the ancient sea bed. The two valves of an oyster shell differ from most other bivalves in

**Above** Jurassic oyster, 3.5 cm wide, carrying the xenomorphic impression of an ammonite that it encrusted. The oyster itself is encrusted by the S-shaped tube of a serpulid worm.

typically being unequal; the left valve, which is cemented to the substrate, is generally larger and often more convex than the right valve, which can be reduced to a lid-like structure. The considerable variations in shell shape found in species of oysters can make it hard to discriminate different species. It is common to find distorted shells resulting from crowding, and shell shape often mirrors that of the surface to which it was cemented. For example, Mesozoic oysters that lived cemented to ammonite shells replicate the shape of the ammonite in negative on the underside of the left valve and in positive on the upper surface of the right valve, a phenomenon called 'xenomorphism'. Xenomorphic oysters may provide the only evidence of the former presence of ammonites in rocks that have suffered loss of aragonitic shells through dissolution.

The size of the cementation area of oysters varies enormously. In some species almost the entire outer surface of the left valve is cemented to the substrate, while other oyster species have a much more restricted area of cementation. Among the latter are species that became cemented to sand grains or tiny shell fragments and adopted a free-lying lifestyle on muddy sea beds. This mode of life was particularly characteristic of Mesozoic oysters such as *Exogyra*, *Texigryphaea* and *Gryphaea*. The evolution of the last of these three genera (see colour fig. 9) has been the subject of several celebrated palaeontological studies. For example, during the early Jurassic successive species of *Gryphaea* exhibit a transition from narrow shells to broad shells that had a larger surface area in contact with the sea bed. The broad shells are thought to have been less prone to overturning by currents, while the greater area in contact with the sediment may have acted like a snowshoe in spreading the weight of the animal and hence preventing the animal from sinking into the mud.

Not far behind oysters in terms of abundance as fossils are the scallop shells or pectinids, yet another group of pteriomorph bivalves. These too usually have two valves of different shapes, one more strongly convex than the other. Pectinid shells generally have strong ribs that radiate from the umbone where there are two small lateral extensions called auricles or ears. Some present-day pectinids (e.g. the Queen Scallop *Chlamys opercularis*) are capable of short bursts of swimming. Used mainly as a means of escaping predators, swimming is achieved by rapid contraction of the adductor muscle (the part of the scallop that figures in human gastronomy), jetting water out from between the valves on either side of the ears. There is something slightly comical about the swimming of a pectinid, the repeated clapping together of the valves is reminiscent of jokeshop false teeth. The

**Above** An inoceramid bivalve, 12 cm long, from the Cretaceous of Hunstanton in Norfolk, England.

likely capacity of fossil pectinids to swim can be judged by the shape of the shell; swimming species tend to have shells with a more acute-angled umbone than non-swimmers. Pectinids are occasionally useful in stratigraphy, as in the Neogene of New Zealand where a succession of distinctive species can be used to recognise rocks of different ages.

Another fascinating, though in this case extinct, group of pteriomorph bivalves are the inoceramids. Sometimes growing 1 metre (3 feet) in length, inoceramids were one of the few animals capable of living directly on the fine-grained coccolith ooze that carpeted the sea bed during deposition of the Cretaceous Chalk in northern Europe. Intact inoceramid shells are not uncommon in the Chalk but are surpassed in abundance by fragments formed after the break-up of their shells. The edges of these shell fragments reveal the distinctive microstructure of the shell that consists of tightly-packed polygonal prisms. Inoceramids as a group met their demise at or very close to the end of the Cretaceous.

Heterodonts, sometimes called eulamellibranchs, are the final major group of bivalves. Most have shells made of aragonite, and the majority burrow into soft sediment on the sea bed. Siphons extend upwards from the buried shell to the sediment surface and are used as conduits to channel water laden with plankton towards the gills. These heterodonts are thus active suspension feeders, or filter feeders. Living heterodonts include such familiar bivalves as cockles, razor shells and clams. Cockles have robust shells with strong ribs. Different species of cockles are found in different parts of the world today, typically living as shallow burrowers in high densities between low and high tide marks. For example, up to 200 individuals of the New Zealand 'cockle' *Chione stutchburyi* can be found per square metre. With such huge populations it is hardly surprising that cockle shells are so routinely found by beachcombers. Razor shells, also intertidal heterodonts, contrast with cockles in having thinner, smoother shells that are greatly elongated. They are deep burrowers and can disappear extremely rapidly into their burrows if disturbed.

Burrowing heterodont bivalves dominate many Cenozoic fossil assemblages, sometimes forming shell pavements or even thick shell beds. Disarticulated shells of dead bivalves usually come to rest in the hydrodynamically most stable attitude, with the outer convex side uppermost and the inner concave side lowermost. This, as might be expected, is the most common orientation in which fossil shells are found. However, the activities of burrowing animals, including other bivalves, worms and crustaceans, may disturb the shells after they have been buried, a process known as bioturbation. Storms can also redeposit dead shells in chaotic orientations. The shells of dead bivalves lying on the sea bed are attractive to a wide range of plants and animals seeking hard surfaces to colonise. Cenozoic fossil bivalves are, for example, often bored by sponges and encrusted by barnacles on their convex surfaces, while the concave surfaces facing the sea bed furnish a protected habitat for more delicate animals such as bryozoans and the tubes of small serpulid worms.

One heterodont remarkable for its size is the giant clam *Tridacna* (see colour fig. 10), a close relative of the humble cockle. The largest shells of *Tridacna gigas* are over 1 metre (3 feet) in size and weigh more than 200 kg (440 pounds). The existence of giant *Tridacna* nestling in coral reefs has led to the entirely fictional spectre of a scuba diver being caught by the leg and held between the tooth-like, zig-zag edges of the closing valves of this huge 'carnivore'. In reality, *Tridacna* feeds on nothing larger than plankton, supplementing its nutrition with energy supplied by symbiotic algae. These zooxanthallae

**Above** Living bivalve *Venerupis*, showing the siphons protruding from between the ends of the valves.

are housed in brightly coloured mantle tissues that extend beyond the gape between the open valves. The fossil record of *Tridacna* extends as far back as the Miocene, but early examples are small compared with the giants of today.

Burrowing into 'soft' sediments like mud and sand using the muscular foot is the most frequent lifestyle adopted by heterodont bivalves. However, a significant number of heterodonts bore into hard substrates such as rocks, corals and wood. Boring is accomplished in many of these species by a combination of mechanical and chemical means. The calcium carbonate of corals and limestones are weakened by acidic secretions, allowing the rotating shell with its rasp-like external ornament to drill more effectively. Boring sometimes ceases when the bivalve reaches maturity, by which time the animal will have excavated a domicile where it is safe from the attentions of predators and protected from destruction by waves. Often the bivalve lines the walls of its borehole with a deposit of calcite for extra security.

Fossil examples of bivalve borings are commonplace. Sometimes only the hole itself remains, in which case it is usually known by the trace fossil name *Gastrochaenolites*. Alternatively, the shell of the bivalve that made the hole can be preserved *in situ*, particularly if the shell had grown too large to be lost through the entrance of the boring following death of the animal. This gives palaeontologists a rare opportunity of pinpointing the exact identity of the maker of a trace fossil. Nonetheless, even in these cases caution must be exercised because some species of bivalves specialise in making their homes in the borings made by other species after the death of the bivalve that made the original excavation.

Several different groups of bivalves have evolved the ability to bore. Two of these deserve further attention. The 'date mussels' or lithophagids often penetrate living corals, maintaining their surface openings during continued growth of the host coral. The linings of the borings made by these symbiotic bivalves show a succession of meniscus-like floors secreted as the bivalve moved upwards to keep pace with coral growth. The second group are the 'ship-worms'. These are not worms at all but teredinid heterodonts whose real identity as bivalve molluscs was unknown until 1733. Infestations of wooden hulled sailing vessels by teredinids presented a major problem to mariners of the past. One teredinid – *Kuphus* – is capable of making holes up to 2 m (6.5 feet) long in wood. Driftwood fossilised in marine environments from the Cretaceous onwards commonly shows the tell-tale signs of attack by teredinids. The long sinuous borings of these animals are sometimes so densely-packed that virtually no wood is left between them. As in modern teredinids, boring into wood was undertaken by extinct species not only to furnish the

**Above** Oblique view of the right valve of the rudist bivalve *Hippurites* from the French Cretaceous, measuring about 20 cm in length.

animal with a secluded home, but also for the cellulose in the wood that could be digested to supplement the nutrition obtained from 'normal' suspension feeding on plankton.

The most bizarre group of extinct heterodonts are the rudists, which have no close living analogues. These animals originated in the late Jurassic and disappeared towards the end of the Cretaceous. Between times, however, they flourished, particularly in the tropical waters of the Tethys, an east-west seaway that extended from east of the current Himalayas to Mexico. Shallow water limestones in southern Europe and the Middle East that were deposited in this ancient ocean often contain fossil rudists in great profusion. Sections of their shells can be seen in many building stones, and individual rudist shells weathered out of limestones have been used to construct walls by French farmers who believed them to be the horns of animals. Many rudist shells do indeed have horn-shaped right valves, up to 1 metre (3 feet) long, capped by a lid-like left valve that is small and can easily pass un-noticed. Such rudists grew upright, typically with the pointed end of the right valve buried in sediment. However, other species of rudists had strongly coiled shells that rested horizontally on the sea bed. The superficial resemblance of rudists to some solitary corals is all the more poignant when taken in conjunction with the fact that many rudists constructed reef-like build-ups and are suspected of having contained symbiotic algae like those found in reef corals. The extreme thickness of rudist shells is one of several pieces of evidence for inferring the presence of symbiotic algae; as with corals, the photosynthetic zooxanthellae may have released carbon dioxide that promoted the growth of the calcium carbonate shells of their hosts.

**Above** Lateral and hinge views of an 8.5 cm wide shell of the bivalve *Arca* from the Pleistocene of Florida.

Not all heterodonts live in the sea. Among freshwater heterodonts are the unionids such as the swan mussel *Anodonta*. A remarkable feature of unionids is that most species have parasitic larval phases, called glochidia, which attach to the gills or fins of fish. Although examples of fossil glochidia can be found in sedimentary rocks deposited in freshwater environments, they have seldom been recognised by palaeontologists. Freshwater bivalves have a long geological history and are particularly well-known in the Coal Measures deposited during the late Carboniferous across wide areas of Europe and North America where genera such as *Carbonicola* flourished.

### *ARCA*: PALEOCENE OR EOCENE–RECENT, WORLDWIDE

The shell of *Arca* has valves that are prolonged posteriorly and have a trapezoidal or almost rectangular outline shape. Between the umbones of the two valves is an extensive ligament area with an ornament of chevrons. Radial ribs and growth lines cover the outer valve surfaces. Internally, the hinge line is long and straight, there are multiple small hinge teeth, and a pallial line lacking a sinus extends between two adductor muscle scars of similar size.

*Arca* lives in shallow water, typically nestling in crevices in rocks or corals to which it is attached by the byssus. There is some doubt as to whether the oldest species of *Arca* is Paleocene or Eocene in age.

### *CRASSOSTREA*: CRETACEOUS–RECENT, WORLDWIDE

The two valves of *Crassostrea* differ significantly in size and shape, the left valve being convex and cemented to a hard substrate, the right valve smaller and relatively flat.

**Above** Collected from Miocene rocks in California, this large shell of the oyster *Crassostrea* is almost 30 cm long.

Some species are large – over 50 cm (20 inches) long – and the shell can be very thick and may contain cavities. The shell surface has a prominent ornament of raised lamellae, sometimes with radial ribs also developed. There are no hinge teeth and only one adductor muscle scar that is kidney-shaped. A deep ligament pit is present in the left valve.

Most modern species of *Crassostrea* inhabit intertidal to shallow subtidal environments and are tolerant of variable salinities as well as turbid water. Two commercially important species of this genus today are the American oyster (*C. virginica*) and the Portuguese oyster (*C. angulata*). In the first of these species, over 100 million eggs may be released by each adult during spawning.

## *GLYCYMERIS*: CRETACEOUS–RECENT, WORLDWIDE

This genus, which includes the living Dog Cockle *Glycymeris glycymeris*, has two identical valves almost circular in outline with small beaks. Subdued ribs

ornament the valve surface and growth lines are also present. The shell margin is crenulated. Internally, there are generally about 6–12 small hinge teeth and a pair of deep adductor muscle scars, almost equal in size, linked by a pallial line without a sinus.

*Glycymeris* inhabits shallow waters in energetic environments, such as submarine sand dunes, making shallow burrows close to the surface of the sea bed.

## *MERCENARIA*: MIOCENE–RECENT, NORTHERN HEMISPHERE

**Above** Interior of a shell of *Glycymeris*, 4.7 cm high, from the Pliocene of eastern England, showing the hinge teeth and muscle scars.

**Above** Interior and exterior views of a 10.5 cm wide fossil shell of the bivalve *Mercenaria*.

This large venerid has thick, equal-sized valves, broadly oval in outline with conspicuous beaks and an external ornament of concentric growth lines. Both valves have three prominent teeth. The pallial sinus is shallow and paired adductor muscle scars are prominent. Valve margins are internally crenulate.

The extant species *Mercenaria mercenaria* is known as the 'quahog' or 'hard shell-clam'. It lives in shelly or stoney mud at shallow depths and is a native of the Atlantic coast of North America, a region in which its fossil relatives may be found in Neogene age deposits.

**Above** Measuring 6.5 cm in width, this Miocene *Pecten* from Virginia is depicted in exterior and interior view.

### *PECTEN*: EOCENE–RECENT, WORLDWIDE

The shell of *Pecten*, usually up to 13 cm (5 inches) in size, comprises a convex right valve and a left valve that can be gently convex, flat or concave. In plan view the shell has the distinctive fan-shape of a scallop. Both valves have 'ears' (auricles). Radial ribs, often flat-topped, form a prominent ornament on the shell surface and produce the crenulate margin of the shell. These are crossed by a concentric ornament of fine lamellae. The hinge line is straight, teeth are lacking, and there is a deep triangular pit for the ligament. A single large adductor muscle scar is present.

*Pecten* itself is one of many genera of scallops (pectinids) which are found commonly as fossils. Their good fossilisation is a reflection of the fact that the shell is predominantly composed of calcite, although they do also have a thin, aragonitic inner shell layer. *Pecten* is a suspension feeder that lives in shallow water. It is neither attached by a byssus nor cemented, but instead has the ability to move by clapping its valves together when disturbed, for example, by a potential predator.

### *PHOLADOMYA*: TRIASSIC–RECENT, WORLDWIDE

The two equal-sized valves of *Pholadomya* are very thin, made of aragonite and usually dissolved away in fossils

**Above** This example of *Pholadomya* is 6.8 cm long and was collected from Upper Jurassic rocks in Wiltshire, England.

that are typically preserved either as steinkerns or with the shell material replaced by crystalline calcite. The valves are strongly asymmetrical, elongated posteriorly towards the large gape in the shell through which the siphon emerged during life. Broad growth lines and widely spaced ribs may be present on the shell surface, as well as an ornament of tiny pustules. Internally, hinge teeth are lacking and the pallial line has a moderately deep and broad sinus.

**Above** A 21 cm tall specimen of *Pinna*, from the Cretaceous of Alberta, Canada, preserved as a steinkern minus the shell material.

There is only one living species of this common fossil genus. This extremely rare animal lives permanently buried, seemingly employing the siphons to process water and organic-rich sediment.

### *PINNA*: CARBONIFEROUS–RECENT, WORLDWIDE

*Pinna* has a medium-sized to very large, up to 80 cm (31 inches), thin shell with equal-sized, elongate valves that are triangular, ham- or wedge-shaped in lateral view. Cross-sections of *Pinna* shells have a characteristic diamond-shape. A long narrow gape along the anterior edge of the shell is for the byssus. Internally the shell is pearly, the anterior adductor muscle scar, situated close to the umbo, is smaller than the posterior muscle scar, which is located about midway along the shell, and there is no pallial line.

*Pinna* lives vertically oriented and partly buried in sediment, anchored by the byssus that attaches to shells and stones, and feeds on plankton in suspension.

### *PLAGIOSTOMA*: TRIASSIC–CRETACEOUS, WORLDWIDE

The shell of this bivalve is thin but individual shells are often large, reaching 20 cm (8 inches) in length. The two valves are mirror images, both being moderately to highly convex and obliquely oval in outline, longer than wide, and having small 'ears' (auricles). A pronounced lunule is present in front of the beaks of each valve. The shell ornament usually comprises fine radial ribs or striae, crossed by occasional subdued concentric growth lines. The hinge line is straight, and teeth are either lacking or limited to one or two broad longitudinal teeth. There is only one muscle scar that is oval in shape and very large.

The pallial line lacks a sinus. It is likely that adults of this genus, which are found commonly in shallow water marine deposits of Mesozoic age, were attached by a byssus during life.

**Above** This shell of *Plagiostoma* from the British Jurassic measures 8.5 cm wide and shows radial ornamentation and growth banding.

## *TEXIGRYPHAEA*: CRETACEOUS OF NORTH AMERICA

*Texigryphaea* is a small to large, up to 11 cm (4 inches), oyster with a strongly arched left valve on which rests a smaller, concave right valve. The left valve has an incurved beak that turns sideways and has a keel running from the beak to the shell margin. Laterally compressed and usually ornamented by concentric growth lines, the shell includes vesicular internal layers unlike the totally solid shell of the related Jurassic genus *Gryphaea*.

During life, the large left valve rested on the muddy sea-floor and the animal was a suspension feeder on plankton.

## *TRIGONIA*: TRIASSIC–CRETACEOUS, WORLDWIDE

The shell of *Trigonia* (see colour fig. 11) has two equal-sized valves that are strongly asymmetrical, trigonal in outline, with a distinct posterior part divided from the

**Above** Oblique view of *Texigryphaea* from the Cretaceous of Texas. A small, lid-like valve articulates with the larger valve, 4.5 cm long, which has a prominent incurved umbo.

rest of the shell by a carina. The anterior part of the shell has a strong wavy ornament almost parallel to the shell margin, the posterior part a perpendicular ornament of ribs bearing nodes. The beak is prominent. Internally, large hinge teeth are visible, the adductor muscle scars are of different sizes, and the pallial line lacks a sinus.

*Trigonia* and related genera, such as *Myophorella* and *Laevitrigonia*, can be common in Mesozoic rocks, typically preserved as steinkerns following dissolution of their aragonitic shells. They were probably shallow burrowers like their living relative *Neotrigonia*.

# Gastropods

The humble garden snail is the tip of the iceberg of a diverse class of molluscs with representatives living in the sea, on the land surface, and in freshwater lakes and rivers. Some 40,000 species of gastropods are alive today and the class has a fossil record extending back to the early Cambrian. The gastropod shell consists of a single valve, although some species have a lid-like operculum that may be calcified to form an additional skeletal element, and other species (slugs and nudibranchs) have lost the shell altogether, at least in the adults. Usually the shell is twisted into a helicospiral coil ending in an aperture through which the muscular foot and head of the snail are extended. Waves of muscular contractions passing along the sole of the foot propel the animal over the surface, generally lubricated by mucus, secreted from the pedal gland, leaving the slime trails that are a tell-tale sign of the passage of terrestrial snails and slugs. A peculiar anatomical feature of gastropods is torsion; during the development of gastropods, usually in the larval stage, most of the soft body and the shell rotate with respect to the foot and head. Always in a counterclockwise direction and varying from 90 to 180 degrees, torsion leads to radical changes in the anatomical structure of gastropods, including twisting of the gut into a U-shape.

The great majority of gastropods, including nearly all marine species, have right-handed shells that coil in a clockwise direction as they grow. If one of these dextral shells is held with the apex uppermost, the aperture will be on the right. The rarer left-handed or sinistral shells coil in an anticlockwise direction during growth; their apertures are on the left when the shell is viewed in this same orientation. Curiously, sinistral shells are more widespread in freshwater than they are in marine gastropods. Two examples of sinistral marine gastropods commonly found as fossils have species names appropriate to their unusual coiling: *Neptunea contraria* and *Busycon contrarium*. It should be emphasised that

shell coiling bears little or no relationship to the phenomenon of torsion mentioned above; whereas torsion is counterclockwise, most gastropod shells are coiled clockwise. However, the oldest apparent gastropods had shells coiled in a single plane (planispiral), lacked torsion, and may in fact have been closer in affinity to a more primitive group of molluscs called monoplacophorans, which are discussed later.

Compositionally, gastropod shells are most often made of the mineral aragonite, making them prone to dissolution during fossilisation. Consequently, it is common to find fossil gastropods preserved as sediment-filled internal moulds (steinkerns), or with the original shell replaced by coarse crystals of calcite. Replacement by silica (silicification) is another mode of preservation that has proved particularly valuable in filling gaps in our knowledge of gastropod fossil history. Some gastropods do, however, secrete partly calcitic shells and these may be found in pristine condition in fossils.

The classification of gastropods is in a state of flux. An older classification dividing gastropods into three main groups has been superseded by a classification recognising five subclasses: Patellogastropoda, Vetigastropoda, Neritopsina, Caenogastropoda and Heterobranchia. Patellogastropods include the true limpets familiar to anyone who has clambered over rocky shorelines at low tide. Limpet shells are mostly made of calcite, have a broad conical shape with a wide aperture, and are generally ornamented by radial ribs and concentric growth lines. The animals feed on algae that grow on rocks, using the batteries of hard teeth forming the radula to scrape off these plants. A characteristic pattern of gouge marks is left on the rock surface by this grazing activity. Fossil examples of such grazing traces made by limpets, as well as by other gastropods and chitons, are known by the trace fossil name *Radulichnus*. Limpets often graze a particular expanse of rock surface while the tide is high before returning to a home site when the sea recedes. A depression in the rock is gradually excavated at the home

**Above** Different species of *Busycon*, both about 8 cm tall, from the Pliocene and Pleistocene of Florida showing opposite coiling – the shell on the left is coiled dextrally, that on the right sinistrally.

site, with the lower rim of the shell fitting tightly around the edge of the depression and forming a perfect seal, circumventing desiccation in the drying air and thwarting attempts by predators (or curious humans) to dislodge the shell.

The second subclass is the Vetigastropoda. When the vetigastropod *Perotrochus quoyanus* was first discovered in the deep sea 150 years ago it was hailed as a living fossil, as nothing similar was previously known alive. Similar 'pleurotomariid' gastropods had long been known in the fossil record; they are especially common in rocks of Mesozoic age but range all the way back to the late Cambrian. These fossil pleurotomariids often inhabited shallow water marine environments, unlike *Perotrochus* and three additional genera that are now known, all living in the deep sea. Pleurotomariids have conical coiled shells with a slit-like indentation in the outer lip of the aperture, giving rise to the common name 'slit shells'. Closure of the slit in older parts of the shell away from the aperture leaves a scar called the selenizone. The pearly appearance of the inner surface of pleurotomariid shells is due to a layer of nacre formed of aragonite crystallites arranged in stacks. Nacre is not unique to pleurotomariids but can be found in a variety of other molluscs including other gastropods, bivalves and cephalopods.

The third subclass, Neritopsina, is relatively minor in importance but includes algal grazing species like the Black rock snail (*Nerita atramentosa*) of New Zealand. Globular in overall shape, shell growth in neritopsines entails rapid whorl expansion, producing an aperture of large area.

The subclass Caenogastropoda encompasses a great many groups familiar to collectors of modern shells, such as the periwinkles, whelks, cone shells, conches, cowries, ceriths, screw shells and volutes. Most of these groups appeared in the Mesozoic and increased in diversity and abundance through the Cenozoic. Although some caenogastropods are algal grazers, others are scavengers, suspension feeders or voracious carnivores. Carnivory is accomplished in several ways. The whelk *Buccinum* uses the edge of its shell to wedge open bivalves and gain access to the flesh within. In two groups – the muricids (rock shells) and naticids (necklace shells) – the ability to bore through the shells of other molluscs, for example oysters, has evolved. Boring is accomplished using the radula as a mechanical rasp, assisted by acidic secretions that soften the shell of the prey. A small circular hole results. This is typically straight-sided when made by muricids but countersunk in naticids. Examples of similar boreholes are particularly common in fossil mollusc shells from the mid Cretaceous onwards. The O-shaped boreholes are given the trace fossil name *Oichnus*. Sometimes shells of the likely culprits are found in the same deposits as *Oichnus,* but in other instances all that remains of the predators is the holes in the shells of their prey.

A more remarkable predatory behaviour is seen in tropical cone shells that literally harpoon their prey. The radula in these snails is shaped like a harpoon tethered on a long proboscis that can be shot out towards a fish,

**Above** Several cemented individuals of the vermetid gastropod *Petaloconchus* from the Pliocene of Florida, each about 3.5 cm in height.

a worm or another gastropod. The unfortunate prey is then tugged towards the snail, which injects it with paralytic venom. Some fish-eating cone shells produce venom sufficiently toxic to kill humans.

Very different from cone shells are the screw shells or turritellids, concentrations of which can be found as fossils in some Cenozoic muddy and silty sediments. Turritellids have high-spired shells with a large number of whorls, reaching up to 18 cm (7 inches) in length. The Cretaceous–Recent gastropod *Turritella* burrows in mud and sand, with the shell at an angle of about 10° to the sea bed, sucking in plankton-laden water. Whereas turritellids lead a relatively inactive life, a related group called vermetids (or worm shells) are unable to move at all. Vermetids become permanently cemented to hard surfaces, constructing mucus nets to capture plankton in much the same way as spiders construct webs to catch flies. The initial parts of vermetid shells closely resemble those of turritellids, but subsequent growth is typically irregular with the shell becoming uncoiled such that successive whorls are not in contact. Both vermetids and *Vermicularia*, another gastropod with a similar shell shape and ecology, may form small reefs when aggregated in large enough numbers.

While many gastropods are themselves predators, they can also fall prey to other animals such as fish, crustaceans and octopus. A principle line of defence against these predators is the shell itself. This can be thickened to make it more difficult to crush, and the vulnerable aperture can be constricted so that predators find it harder to gain access to the flesh of the animal within. Another defensive strategy is to grow a high spired shell into which the body can retract deeply. Features of the gastropod shell that deter predation have become increasingly common through geological time: many Palaeozoic gastropods have plain shells with large apertures, whereas Cenozoic and Recent shells are often ornamented with spines or flanges and have apertures that are constricted, as in the cowries such as *Cypraea*. This trend has paralleled the progressive appearance and increasing abundance of predators in the fossil record and it has been argued that it indicates an escalating arms race between predators and prey. Not all attacks by the predators of gastropods meet with success. Often in fossil gastropods it is possible to see evidence of failed predation, for example, stepped growth checks with jagged edges that record the trauma in the life of animals attacked via the shell aperture.

Camouflage is another method employed by gastropods in their attempt to prevent becoming victims of predators. This is taken to an extreme degree in the remarkable carrier shell, *Xenophora*. Carrier shells deliberately cement foreign objects to the outer edges of their shells, especially the shells of other molluscs and pieces of coral. Sometimes these objects fall off during

fossilisation, leaving only scallop-like depressions in the shell to show where they were once attached.

Some caenogastropods attain a considerable size, especially an extant genus called *Campanile* which includes a gigantic Eocene species growing more than 50 cm (20 inches) in length. At the opposite end of the size spectrum, there are huge numbers of tiny gastropod species with shells measuring only a few millimetres in size and very easily overlooked by zoologists and palaeontologists alike.

The final gastropod subclass is the Heterobranchia. The two principal groups of heterobranchs are the opisthobranchs and pulmonates. The Opisthobranchiata is a superorder of mostly marine snails including some species that have lost their shells during evolution, others

**Above** Pleistocene specimen of the cowrie *Cypraea* showing the narrow, toothed aperture extending the full height (6.7 cm) of the shell.

with shells that begin by coiling sinistrally but then switch to a more conventional dextral coiling, and a few bizarre species with shells comprising two mirror image valves that can easily be mistaken for bivalve molluscs. The nudibranchs or sea-slugs are without doubt the most visually striking opisthobranchs. These brightly-coloured animals, lacking shells and therefore not found as fossils, often have feathery respiratory organs (cerata) on the back of the bilaterally symmetrical body, and many species are capable of swimming, gliding gracefully through the water. Among shelled opithobranchs found as fossils are the Bullidae or bubble shells that range back to the Jurassic and today are burrowers in muddy sand.

The Nerinoidea are a distinctive group of Jurassic and Cretaceous opisthobranchs with high-spired shells.

**Above** Jurassic *Aptyxiella*, 3.2 cm high, preserved as an internal mould after infilling by hardened lime mud and dissolution of the shell.

The interior chamber of the shell is greatly reduced in advanced species of nerineids by the growth of unusual ridge-like internal thickenings of shell material. Most nerineids probably burrowed through lime-rich mud, ingesting the mud as they went to feed on organic matter, in the manner of an earthworm processing soil. The shell thickenings are believed to have functioned as repositories for the excess calcium carbonate taken in during deposit feeding. Nerineacean fossils are commonly preserved as moulds following dissolution of the aragonitic shell. Internal moulds (sediment casts or steinkerns) of the nerineacean *Aptyxiella* found in Jurassic rocks of southern UK resemble screws and are known as 'Portland Screws' in folklore.

Many terrestrial snails, including the common garden snail *Helix* and slugs, as well as freshwater snails, belong in the main to the Superorder Pulmonata. Pulmonates differ from other gastropods in lacking gills and respire using lungs formed from the mantle cavity which is kept moist and is provisioned with a liberal blood supply for absorbing oxygen. It is this feature that has allowed pulmonates to invade terrestrial habitats with great success; today some 15,000 species of pulmonates live on the land. Pulmonate shells are usually thin and more fragile than shells of marine gastropods that have better access to calcium in the sea water they inhabit. Fossils of terrestrial pulmonates are seldom found, although their shells can occur in sand dunes and deposits of loess (wind blown silt). However, freshwater pulmonates can be found as fossils in ancient lake deposits. They include the various species of the pond snail *Lymnaea*. The oldest pulmonates apparently date from the Carboniferous, but terrestrial species did not appear in any significant abundance until the Cretaceous.

*CAMPANILE*: CRETACEOUS–RECENT, WORLDWIDE

This giant marine gastropod has a thick, high-spired, turretted shell which can reach 60 cm (23 inches) in length and comprise more than 30 whorls. Sectioned

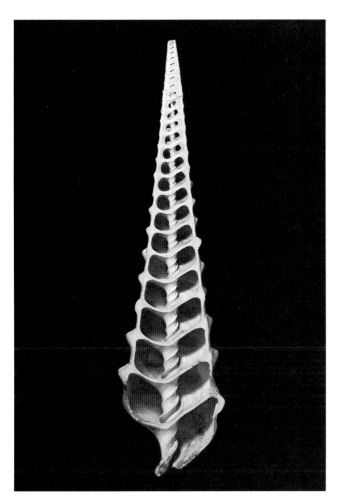

**Above** The central columella is conspicuous in this vertically-sectioned specimen of the gigantic French Eocene gastropod *Campanile* which measures 40 cm in height.

## *LOXONEMA*: ORDOVICIAN–CARBONIFEROUS, WORLDWIDE

A high-spired gastropod with a relatively smooth shell, *Loxonema* has numerous convex whorls separated by deep sutures. Sinuous growth lines may be present. The aperture is long and ovoidal, the operculum unknown. A deep notch occurs in the outer lip of the aperture.

This genus is moderately common in the Silurian, typically preserved as a mould left after dissolution of the aragonitic shell. As in Recent slit-shells, the notch on the outer lip of the aperture was probably an exhalent channel allowing disposal of waste material.

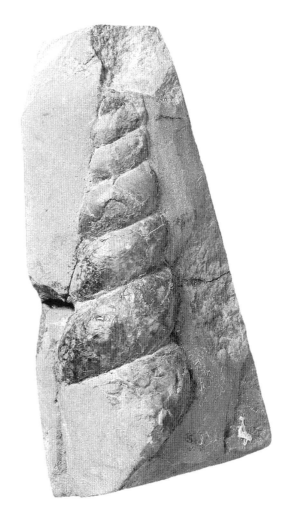

**Above** A British Silurian specimen of *Loxonema*, 3.5 cm long, partly developed from the enclosing sediment.

specimens show an infilling of the older whorls by secondary calcite, plus the presence of two spiral ridges on the columella that forms the central pillar of the shell. External ornament comprises spiral cords, a prominent row of blunt knobs, and fine, curved growth lines. The aperture is spindle-like, inclined obliquely to the long axis of the shell, and separated into siphonal channels. Both inner and outer lips are thickened, the former by a callus.

Only one species of this caenogastropod genus is alive today. This feeds on algae in shallow, warm seas, a lifestyle also likely to have pertained in the extinct fossil species.

**Above** This example of the thin-shelled, freshwater gastropod *Lymnaea*, measuring 3.7 cm in length, was collected from the Eocene of southern England.

## *LYMNAEA*: JURASSIC–RECENT, WORLDWIDE

The thin shell of *Lymnaea* is dextral and has a large body wall and slender, pointed spire which accounts for about half of the total shell height. The whorls are slightly rounded and the aperture is oval with a plain outer lip and inner lip having a thin callus. Shell ornament is subdued, although growth lines or spirals may be present.

This pulmonate gastropod lives in freshwater, grazing on plants but occasionally consuming small invertebrates and decaying matter. A well-known Recent species is *L. stagnalis*, otherwise called the Great Pond Snail.

## *NATICA*: PALEOCENE–RECENT, WORLDWIDE

The shell of the caenogastropod *Natica* is bulbous, almost spherical, with a large final (body) whorl and a small, low spire. An ornament of growth lines may be developed. The aperture is wide and semicircular in shape, equipped with a calcified operculum during life. The inner lip is thickened by a callus that may partly plug the wide umbilicus.

*Natica* lives in sand and silt where it preys on bivalve and gastropod molluscs. This is accomplished by first drilling a hole in the shell of the prey and then removing the fleshy tissue within using the proboscis. Holes made by *Natica* are about 1 millimetre in diameter and countersunk. They fall within the trace fossil species *Oichnus paraboloides* and can sometimes attest to the past presence of *Natica* (or related genera) even when shells of the gastropod are lacking.

**Above** Preserving the original pigmented spots, this Miocene specimen of the predatory gastropod *Natica* is almost 3 cm in width.

## *NEPTUNEA*: EOCENE–RECENT, NORTHERN HEMISPHERE

This gastropod is moderately large, up to about 10 cm (4 inches), fusiform and thick-shelled. The convex whorls are separated by deep sutures. Shell ornament comprises

**Above** Sinistrally-coiled shell of *Neptunea contraria*, 8 cm high, from the Pleistocene of Suffolk, England, where this species is a distinctive fossil in the Red Crag Formation.

fine spiral threads and prominent growth lines. The aperture is oval and prolonged into a broad but moderately short siphonal canal, the outer lip smooth and inner lip with a callus.

While most species of this caenogastropod are conventionally dextral, one notable species is sinistral, *Neptunea contraria*, a characteristic fossil of the Pliocene Red Crag of East Anglia, England. *Neptunea* is a whelk that occurs in relatively cold seas today, from shelf to abyssal depth. It is a scavenger or carnivore, preying on bivalve molluscs and scavenging fish corpses.

### *PLEUROTOMARIA*: JURASSIC–CRETACEOUS, WORLDWIDE

This vetigastropod has a large, often up to 8 cm (3 inches), conical shell (see colour fig. 12) with convex whorls separated by sutures. The aperture is rectangular in shape and has an outer notch, the former position of which is shown as a spiral band (selenizone) visible in the older parts of the shell. The thick shell comprises inner aragonitic and outer calcitic layers. An ornament is present on the shell surface of strong spiral threads intersected by oblique lines that may thicken to form tubercles on the shoulders of the whorls.

Modern relatives of this 'slit shell' inhabit deep water marine environments, but this was not true for the Mesozoic genus *Pleurotomaria,* which is found in shallow water deposits.

### *STRAPOROLLUS*: DEVONIAN–PERMIAN, WORLDWIDE

The shell of *Straporollus* (see colour fig. 13) at first appears to be almost planispiral, but closer inspection reveals that one side is raised into a low spire and the other is depressed. Whorls are ridged and ornamented by transverse growth lines. Older parts of the shell cavity are sealed off by a succession of septa-like walls that can be seen when specimens are sectioned. This late Palaeozoic genus was probably an algal grazer.

### *STROMBUS*: MIOCENE–RECENT, WORLDWIDE

A medium to large marine caenogastropod, *Strombus* has a thick fusiform shell without an umbilicus. Strong nodes or spines may be developed at the shoulders of the whorls and growth lines are also present. The aperture is long and narrow, with a thickened outer lip in mature shells that is expanded and wing-like, and an inner lip with a smooth callus. The small, horny operculum is not fossilised.

Modern shells of the conch *Strombus* are familiar bathroom ornaments, especially the Queen Conch

**Above** Pliocene specimen, 6.3 cm high, of *Strombus* from Florida.

(*S. gigas*), an edible Caribbean species that can grow up to 30 cm (12 inches) in size. Living animals are slow-moving, relying on the thick shell for protection against predators, and are herbivorous.

## *TURRITELLA*: CRETACEOUS–RECENT, WORLDWIDE

This marine caenogastropod has a high-spired, turreted shell which is long and slender, exceeding 10 cm (4 inches) in some species. Regular spiral cords, some ridge-like, and growth lines ornament the external surface of the shell. The rounded quadrate aperture has a plain outer lip and an inner lip with a thin callus. The chitinous operculum is not preserved in fossils

Living species of *Turritella* – known as auger shells – burrow in mud, silt and sand. They feed both on the sediment and by producing ciliary currents or mucus strings that entrain food particles. It is common to find fossil *Turritella* shells in clusters, possibly reflecting the aggregative tendencies of the living snails, although dead shells may also be swept into drifts by tidal currents.

**Above** This 9.5 cm tall Eocene specimen of *Turritella* from the Paris Basin contains two circular borings of the type made by predatory gastropods and given the trace fossil name *Oichnus*.

## *VOLUTOSPINA*: PALEOCENE–OLIGOCENE, WORLDWIDE

The shell of this caenogastropod genus is broadly spindle-shaped with a moderately short spire, up to one-quarter of the total shell length, and a large body whorl. The elongate aperture tapers towards an open siphonal channel, and the inner lip has a thin, smooth callus. Strong axial ribs ornament the shell surface. On the shoulders of the whorls these ribs are prolonged into thorn-like spines. Spiral bands may also be developed as additional shell ornament.

*Volutospina* is a characteristic Paleogene marine gastropod, typically found in fine-grained sediments deposited in shallow water environments. Like extant related volutes, it was probably a predator and used a protrusible proboscis to feed on other molluscs.

**Above** An Eocene shell of *Volutospina* measuring 5 cm and showing prominent spines on the shoulders of the whorls.

# Cephalopods

Ammonites are perhaps the most popular of all fossils. Treasured by collectors, their beautiful spiral shells are readily identified as fossils by almost anyone who has ever heard of the word fossil. Ammonites are extinct representatives of the molluscan class Cephalopoda. While they are undoubtedly the best-known fossil cephalopods, they are by no means the only cephalopods as this class also includes the extant octopus, squid, cuttlefish and nautilus, as well as the extinct belemnites.

Regarded as the most advanced molluscs, cephalopods have a sizeable head with arms and/or tentacles surrounding a beaked mouth. Their large and complex eyes show remarkable similarities to those of vertebrates despite having an entirely separate evolutionary origin. The wily behaviour of the octopus betrays the generally high levels of intelligence shown by cephalopods. Vision and intelligence are necessary for their active lifestyles; most are fast-moving predators that swim using a form of jet propulsion, shooting water through a muscular funnel called the siphon. Many cephalopods can also squirt clouds of ink to hide their position from enemies, or change the colour of their bodies to fit in with that of their surroundings. Primitively, cephalopods have external shells of calcium carbonate, like those seen in other classes of molluscs. However, a key distinguishing feature of the cephalopod shell is that it is divided into chambers by partitions called septa. During the evolution of cephalopods the shell has been internalised in many groups, as in squid and cuttlefish, or lost altogether, as in the octopus.

About 650 species of cephalopods live today. They range from tiny species of octopus just a few millimetres in size, to the fabled giant squid – *Architeuthis* – which grows up to 18 m (59 feet) in length. All cephalopods are marine animals, none can be found in freshwater. The overwhelming majority of extant species belong to a subclass called the Coleoidea that includes cuttlefish, squids, vampire squids and octopuses. Coleoids have either a reduced shell or no shell at all. Consequently, they are not particularly common as fossils, with the exception of the extinct belemnites that had robust, bullet-shaped internal shells largely made of the resistant mineral calcite. Two other subclasses – Nautiloidea and Ammonoidea – contain the majority of cephalopods found as fossils. These will therefore be emphasised here.

## NAUTILOIDS

A mere five or six species of nautiloids live today and these belong to only two genera, *Nautilus* and *Allonautilus*. They represent the tip of the iceberg of a primitive cephalopod group much more diverse in the geological past with almost 1000 fossil genera and making its first appearance in the late Cambrian.

Shells of the living nautilus are common ornamental curios, frequently cut in half to reveal their pearly interiors and chambered internal structure. The existence of these chambers, partitioned from one another by curved septa and interconnected by a tubular structure called the siphuncle, is crucial to an understanding of how nautilus lives. When the animal is alive the chambers are filled partly with gas and partly with liquid. The animal is able to adjust the balance of gas to liquid in each chamber, thereby controlling buoyancy; by introducing more gas, the animal ascends in the water column, by introducing more liquid it sinks. The main body of nautiloids is located in the living chamber, the space between the last formed septum and the aperture of the shell. In the modern *Nautilus* there are up to 90 tentacles (or arms), primitive eyes without lenses, two pairs of gills, gonads, and gut. When the tentacles are retracted, a tough hood covers the aperture and protects the animal within. There is also a structure called the hyponome through which water is ejected to propel the animal backwards. Compared with other living cephalopods, *Nautilus* is a poor swimmer and animals kept in tanks fail to exhibit the same degree of intelligent behaviour seen in other cephalopods. Living populations of *Nautilus* are found in the Indo-Pacific at depths down to about 500 metres. They are

predators, capturing crustaceans and fish using their tentacles and slicing them up with a horny beak (fossil examples of nautiloid beaks are occasionally found, going by the name rhyncholites). Individuals take more than five years to reach sexual maturity and produce a mere dozen or so eggs a year.

Not only were nautiloids much more diverse in the geological past, but they also exhibit a considerably greater range of shell shapes. Some, particularly in the Mesozoic and Cenozoic, have tightly-coiled shells similar to the shells of the modern *Nautilus*, but others are very different. Particularly common in the Palaeozoic are nautiloids with straight shells, increasing in diameter gradually towards the aperture. These are often known informally as orthocones, a term referring to the straight, uncoiled shell. Examples of orthocones can often be seen in limestones employed as building or paving slabs, for example, the Orthoceras Limestone used in the steps of St Paul's Cathedral in London. This Ordovician limestone was quarried on the Swedish island of Öland. Cut and polished Moroccan limestones also commonly feature orthocones, while in China these fossils are known as Pagoda Stones on account of their resemblance with the stepped towers found in Buddhist temples. It is believed that most orthocones swam with their shells held horizontally. This orientation makes sense by analogy with modern squid whose elongate bodies also swim in this attitude, and is supported in some orthocones by the existence of traces of colour patterning confined to the upper (dorsal) surface, as in modern free-swimming cephalopods and fish. Many orthocone and other Palaeozoic nautiloids secreted so-called 'cameral deposits' within the chambers of the shell. Cameral deposits are often found on one side only, forming a kind of ballast that not only kept the dorsal surface of the animal uppermost, but also perhaps acted as a counterbalance preventing the weight of the main body tipping the animal vertically.

Other shell shapes found among fossil nautiloids include gently curved (cyrtocones), openly coiled

Nautiloids were probably the top predators in the sea for much of the early Palaeozoic. At almost 10 m (33 feet) long, the largest Ordovician examples are the biggest invertebrate animals known from the Palaeozoic and must have been truly fearsome creatures. As mentioned above, nautiloids are important fossils in many Ordovician and Silurian limestones, especially slowly deposited pelagic limestones formed beyond the range of terrestrially derived sediment. However, by the Devonian, nautiloids were beginning to decline in both abundance and diversity as the ammonoids proliferated. A major crisis in nautiloid evolution occurred at the end of the Triassic when the group narrowly survived total extinction. Nautiloids often occur in small numbers in Jurassic and Cretaceous rocks dominated by ammonites, and a few genera lingered on into the Cenozoic, experiencing another near fatal extinction in the Miocene. Today's nautiloids are vulnerable to extinction through human activities, especially the collection and sale of their beautiful shells as curios.

## AMMONOIDS

The most emblematic of all fossils, ammonoids, first appeared in the early Devonian and persisted until the very end of the Cretaceous. Their evolution is characterised by a 'boom and bust' pattern; groups of ammonoids diversified only to be decimated by extinctions that left just a few survivors to initiate the next evolutionary radiation. Among the subgroups of ammonoids that flourished at particular times are the Goniatitina (goniatites) of the Carboniferous and Permian, the Ceratitina (ceratites) of the Permian and Triassic, and the Ammonitina (ammonites proper) of the Triassic, Jurassic and Cretaceous. The rapid evolution of ammonites has made them paramount among all macrofossils for the purposes of stratigraphical zonation, particularly in the Jurassic.

It is believed that ammonoids evolved from nautiloid ancestors called the Bactritoidea. These animals are unusual among nautiloids in having the siphuncle close

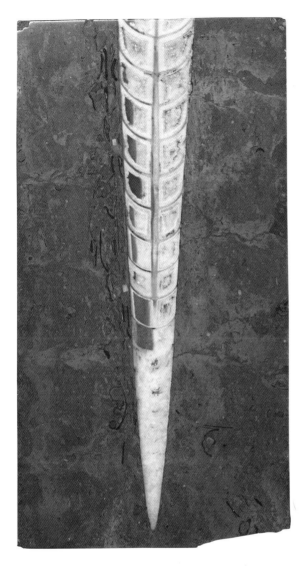

**Above** Polished slab of Chinese Ordovician limestone containing a 20 cm long section of an orthoconic nautiloid, popularly known as a Pagoda Stone.

(gyrocones) and bulbous, spindle-like forms (brevicones). Particularly interesting are species – oncocerids – having brevicone shells that are often combined with a peculiar constriction of the aperture so that the opening comprises a few narrow lobes. Unlike other nautiloids, these Ordovician–Silurian oncocerids possibly lived very close to the sea bed with their shells oriented vertically. Tentacles protruding from the aperture may been employed to grub around for food.

to the outer margin (venter) of the shell, a feature also seen in most ammonoids and contrasting with the more central positioning of the siphuncle that typifies nautiloids. Bactritoids in addition have a bulbous ammonoid-like protoconch, the first-formed part of the shell. They lack the cameral deposits found in many other nautiloids and it is thought that bactritoids with straight, orthoconic shells maintained a near vertical life orientation in contrast to the horizontal orientation of other orthocones discussed previously.

The most striking difference between advanced ammonoids and nautiloids is undoubtedly found in the suture lines. These are the traces where the septa that partition the shell into chambers meet the inner surface of the shell itself. Suture lines are often clearly visible in fossil ammonoids because the aragonite shell is prone to dissolution, exposing the sediment- or crystal-filled chambers outlined by suture lines. Sutures in nautiloids are simple and gently curved; those of ammonoids exhibit varying degrees of elaboration. In the goniatite ammonoids, sutures are typically zig-zag, ceratites have lobes (inflections facing the shell origin) that are frilled but intervening saddles (inflections facing the shell aperture) that are plain, whereas ammonites have both lobes and saddles bearing complex frills. The reason why ammonites evolved such complex sutures has long been a topic of debate among palaeontologists and is still not fully resolved. A popular idea is that sutural complexity strengthens the shell. One of the problems faced by cephalopods with shells containing gas-filled chambers is the increasing water pressure exerted on the shell as animals descend into ever deeper waters, pressure that may ultimately cause the shell to fracture and implode. Perhaps more important, however, is water pressure forcing the main body of the animal against the final septum behind which is a gas-filled chamber. As this pressure is transmitted to the load-bearing walls of the shell along the suture lines, longer suture lines, made possible by complex folding, could provide the greater structural strength needed to prevent catastrophic failure. An entirely different hypothesis for the function of complex sutures is concerned with storage of liquid within the shell chambers. In living nautilus, septa are covered with a layer of mucus that retains the liquid crucial to the adjustment of buoyancy. The larger surface area of the fluted

**Above** From the Devonian of Germany, this specimen of the ammonoid ancestor *Bactrites* measures 5 cm in length.

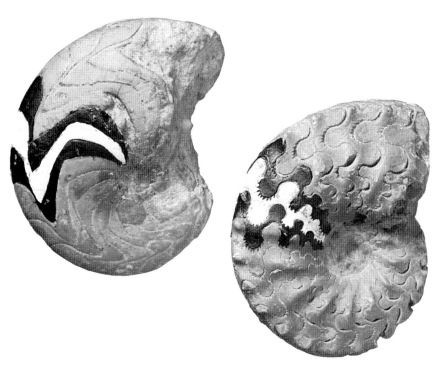

**Above** Ammonoids painted to show different types of suture lines – goniatitic (left), ceratitic (centre, in *Ceratites*) and ammonitic (right, in *Phylloceras*).

ammonite septa would have allowed a greater volume of this liquid to be stored.

Variations in the septal patterns provide a kind of fingerprint for identifying ammonoid species. Ammonoids also exhibit substantial differences in shell shape, useful in distinguishing between the myriad of different species that evolved during their 300 million year history. Among ammonoids with shells coiled in one plane (planispiral), the coiling varies from loose (evolute), with successive whorls barely overlapping, to tight (involute), with new whorls overgrowing and covering most of the surface of the preceding whorl. Viewed facing the aperture, shells can range from very narrow and almost discus-like, to broad and globular. The shell surface in some species is nearly smooth but it may instead bear fine growth lines, regularly spaced ribs or hummock-like tubercles. A keel or alternatively a groove is sometimes developed around the outer edge. Deviations from conventional coiling occur in the 'heteromorph' ammonites particularly characteristic of the Cretaceous. Some heteromorph species began

growth as conventionally coiled shells but became progressively uncoiled during later growth. Other heteromorphs have straight or hooked shells, and yet others have helicospirally coiled shells resembling turritellid gastropods. The bizarre Japanese heteromorph *Nipponites* has a shell that looks like a random tangle of whorls, so irregular is the coiling. Loosely coiled ammonoid shells are also found among a small Palaeozoic group called the agoniatites, thought to be direct descendants of the bactritoid nautiloids from which ammonoids arose.

The biology of ammonoids is, unfortunately, far less well understood than is the distribution of ammonoid species in time and space. Most attempts to interpret ammonoid biology begin by drawing analogies with the living nautilus. However, ammonoids not only exhibit a number of significant differences in shell morphology from living nautilus, but as a group also display a far greater variation in shell shape, placing question marks over how far analogies can be taken. Variation in shell shape suggests that ammonoids encompassed a range of different ecologies: some species probably moved around in a sluggish fashion close to the sea bed, whereas others with more streamlined shells were likely to have been rapidly-swimming predators of the open oceans. Gut contents have seldom been found associated with ammonoids, but those that are known indicate that ammonoids were carnivorous, as are all present-day cephalopods. Some species of ammonites have aptychi, a pair of shield-like plates of horny material or calcite generally found dissociated with the rest of the shell, but occasionally preserved *in situ* close to the aperture. While many palaeontologists believe that aptychi served to close the aperture when the tentacles were retracted, like the hood of the living nautilus, others have suggested that they formed part of a jaw mechanism and were used to shred food.

Sexual dimorphism has been inferred for many species of ammonoids, although it may not have been universal. Co-occurring pairs of 'species' with identical early growth stages but contrasting in later growth are thought to represent individuals belonging to the two sexes. Generally one of these morphs, the macroconch, is appreciably larger than the other morph, the microconch (see colour fig. 15). This is not simply due to age differences; it can be shown that both morphs were adults by the occurrence in them of closely spaced septa, and often of extensions called lappets around the aperture in the microconchs, both features indicating maturity in ammonoid shells. Although there is no way of knowing for certain, it is thought that the macroconchs were females and the microconchs males. Strangely, beds rich in ammonites often show an overwhelming dominance of either microconchs or macroconchs, and a lack of juvenile shells. Segregation of the sexes and absence of young might indicate that the site of burial and fossilisation was not the same as the place where breeding populations actually spent most of their time.

While it is tempting to assume that ammonoids had similar life histories to the living nautilus – growing slowly, taking a long time to reach maturity, and producing a small number of eggs each year – there are reasons to believe that this was not so. The shell of a newly hatched nautilus larvae is large, about 2 cm (1 inch) in diameter, whereas that of ammonoids was much smaller, generally less than 1 millimetre. It seems likely that ammonoids produced large numbers of small eggs rather than small numbers of large eggs. In this respect they are more comparable with modern coleoids than with nautilus, and perhaps also shared the rapid growth rates and short life spans typical of coleoids.

Ammonoids are typically less than 20 cm (8 inches) in diameter but a significant number of species are much larger. A late Jurassic ammonite appropriately named *Titanites* could grow to almost 1 metre (3 feet) in diameter but even this is small compared to the largest ammonites from the Cretaceous, which are up to 3 m (10 feet) across. Gigantism among animals in general is not only associated with overpowering and eating large items of prey, but can also be an effective means of not falling

**Above** Paired aptychi visible close to the aperture of a poorly-preserved 5.4 cm diameter shell of the Jurassic ammonite *Oppelia* from Germany.

prey to other predators. Ammonite shells have been found punctured by holes matching the teeth of mosasaurs, marine reptiles that lived at the same time as ammonites, but a 3 m (10 feet) diameter ammonite may have been too much for a mosasaur to tackle.

As previously noted, ammonoid shells are made of aragonite. In ammonites this comprises three layers, a thick layer of nacre sandwiched between thinner layers of prismatic crystals. Nacre has minute plate-like crystals arranged parallel to the shell surface. Ammonites preserving nacreous aragonite often have an iridescent, mother-of-pearl lustre. This represents one end of the spectrum of ammonite preservation. At the opposite extreme are ammonites in which no shell material at all remains and the individual chambers have become separated from one another. Such fossils have a peculiar appearance, shaped like bent discs with fluted edges that represent moulds of the frilled septa. Ammonoids in clay or shale are often crushed and the best specimens in such sediments may be found in concretions where early cementation prevented compaction. The effort involved in splitting hard concretions parallel with bedding can be richly rewarded by the discovery of exquisite ammonoids preserved in three dimensions. Ammonoids recovered from shales and clays often glint with a patina of gold-coloured pyrite. Pyritisation was mediated by sulphur-reducing bacteria acting on decaying organic matter.

## ASTEROCERAS: EARLY JURASSIC OF EUROPE AND NORTH AMERICA

An often large ammonite, *Asteroceras* has a broad umbilicus (the concave inner part of shell between the outer whorls), high whorls that enlarge rapidly, and strong, smooth ribs curving forwards. A keel is present around the outer edge of the shell, flanked by deep grooves that may be lost on the outer whorl.

Specimens of this ammonite from the coast of Dorset in southern England often have their sealed inner

chambers filled with transluscent brown calcite that was precipitated after burial. In contrast, the body chamber, open at the shell aperture, is typically filled by fine-grained sediment.

**Above** A 16 cm diameter specimen of the Jurassic ammonite *Asteroceras* from the Lower Jurassic of Dorset, England.

## BACULITES: LATE CRETACEOUS, WORLDWIDE

This ammonite has a straight shell, except for a coiled initial part that is often lost in fossils. The shell can be over 1 metre (3 feet) long in extreme examples. In cross-section the shell is oval, the surface either smooth or bearing an ornament of striations or weak ribs. Suture lines are florid.

**Above** Part of a steinkern of *Baculites*, measuring 8 cm long and showing intricate suture lines, from the Cretaceous of Colorado.

Some late Cretaceous deposits are packed with *Baculites* and it is possible that these represent schools of cephalopods that fed on plankton. This genus is among the last of the ammonites to have become extinct at the end of the Cretaceous.

## *CERATITES*: TRIASSIC, NORTHERN HEMISPHERE

The shell of *Ceratites* is robust, moderately evolute and has a broad umbilicus (see p. 89). Strong, widely spaced ribs end in ventral tubercles. The body chamber is short and the sutures are typically ceratitic, with digitate lobes directed towards the shell origin, and smooth, rounded saddles directed towards the shell aperture.

A characteristic Triassic ammonoid, *Ceratites* is particularly common in a stratigraphical formation called the Muschelkalk ('mussel limestone') which is developed across continental Europe.

## *DACTYLIOCERAS*: EARLY JURASSIC, WORLDWIDE

*Dactylioceras* has a markedly evolute shell in which successive whorls overlap only very slightly. A large number of whorls are therefore visible and the shell is described as serpenticone. In cross-section, the whorls are circular, subcircular or slightly compressed. There are numerous closely spaced ribs that are straight on the side of the whorl but often bifurcate over the venter. The body chamber is long.

Perhaps more than any other ammonite, *Dactylioceras* resembles a tightly coiled snake. Indeed, specimens from Whitby in northern England were once believed in folklore to be snakes turned to stone by the abbess St Hilda when seeking to clear a site for monastic buildings.

**Above** *Dactylioceras*, represented here by a 9.5 cm wide individual, is a characteristic and abundant ammonite in the Lower Jurassic of north Yorkshire, England.

**Above** In marked contrast to *Dactylioceras*, this Carboniferous *Goniatites* has a shell in which successive whorls overlap strongly, giving a narrow, deep umbilicus.

## *GONIATITES*: CARBONIFEROUS, WORLDWIDE

The globular, almost spherical, shell of *Goniatites* is strongly involute, with successive whorls overlapping greatly, resulting in a small umbilicus. Fine growth lines or ribs may ornament the thin shell, which is otherwise very smooth. The goniatitic suture lines have a zig-zag pattern, the lobes being pointed and the first saddle angular or subangular.

The broad, poorly streamlined shell makes it likely that of *Goniatites* was a slow swimmer. It is often found associated with reef facies in the early Carboniferous and it is possible to envisage the living animal cruising around the shallow reef in search of food.

## *HOPLITES*: EARLY CRETACEOUS OF NORTH AMERICA AND EUROPE

The shell of *Hoplites* has a moderate umbilicus and bears prominent ribs that branch from strong umbilical tubercles and curve forwards, terminating at equally strong ventrolateral tubercles. A sulcus is present around the outside of the shell (venter) which interrupts the passage of the ribs. Whorls are rectangular or trapezoidal in section.

Species of this genus and closely related genera have proved to be immensely useful in stratigraphical correlation over the region of their occurrence. The genus name *Hoplites* is derived from heavily armed fighting men of ancient Greece, the Hoplites.

**Above** *Hoplites*, a strongly-ribbed Cretaceous ammonite. This 7.8 cm wide specimen is from the Cretaceous of southern England.

## *PHYLLOCERAS*: JURASSIC–CRETACEOUS, WORLDWIDE

The shell of this ammonite is involute with a small umbilicus (see p. 89). Whorls are oval in section. Ornament is subdued, usually comprising only fine growth lines or gentle folds. Suture lines have numerous saddles and lobes, the saddles being characterised by their rounded, spatulate endings.

*Phylloceras* gives its name to the Order Phylloceratida, which played a pivotal role in ammonoid evolution. This order survived when other ammonoids (e.g. ceratites) became extinct during the late Triassic, subsequently to diversify and spawn the ammonites *sensu-stricto* of the Jurassic and Cretaceous.

## *PLACENTICERAS*: CRETACEOUS, WORLDWIDE

An involute ammonite with a discus-like shell, *Placenticeras* grew up to 50 cm (20 inches) in diameter. The umbilicus is small and ribs are lacking or very faint. The outer edge of the shell (venter) is flattened.

The streamlined shape of *Placenticeras* points to a capacity for rapid swimming, either to assist in the

capture of prey or to escape from being eaten itself. Nonetheless, puncture marks in the shells of some individuals have been interpreted as bite marks made by predatory marine reptiles.

## *SCAPHITES*: CRETACEOUS, WORLDWIDE

*Scaphites* (see colour fig. 17) is a small ammonite in which the tightly coiled inner whorls are followed by a living chamber comprising a straight shaft and a curved hook. The aperture is constricted, often has a thickened collar, and may have lappets. An ornament of bifurcating ribs is present, some leading to umbilical and/or ventral tubercles.

It is thought that *Scaphites* was a sluggish swimmer that lived close to the sea bed.

## COLEOIDS

As already mentioned, the great majority of living cephalopods belong to the Coleoidea, a group that includes the cuttlefish (sepioids), squids (teuthoids), octopuses (octopods) and vampire squids (vampyromorphs). They are of great importance in today's oceans; indeed, some scientists believe that their total biomass may be as high as that of fish. The shells of modern coleoids, if present at all, are reduced and internal, enveloped by soft tissues. This does not bode well for their prospects of fossilisation. However, one extinct group of coleoids – the belemnites – secreted robust internal shells that have left an excellent fossil record, while some other coleoids provide spectacular examples of soft-bodied preservation through mineralisation of the soft tissues. The earliest fossil coleoids are from the early Devonian.

Modern coleoids have either eight or ten tentacles, one pair of gills, complex nervous systems, elaborate eyes with lenses, and pigment cells called chromatophores in their skin that permit the animals to mimic the background for camouflage and also to engage in colourful behavioural rituals. The ink sac is another key feature of coleoids. This contains a fluid,

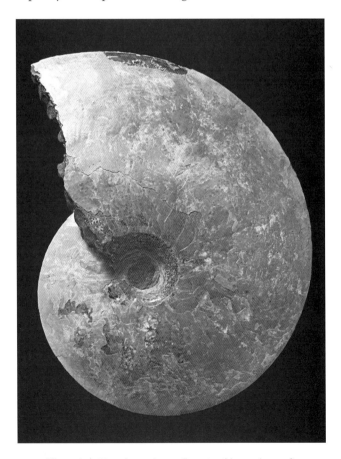

**Above** Measuring 15 cm in maximum diameter, this specimen of *Placenticeras* from the Cretaceous of Dakota has a smooth shell quite unlike that of *Hoplites* (see p. 93).

consisting of the pigment melanin plus mucus, that can be ejected out of the animal to form a black or brown cloud in the sea water, producing a kind of 'smoke screen' mirroring the shape of the body, confusing potential predators and allowing the coleoid to make its escape. Amazingly, dark patches representing remnants of ink sacs have been described in several fossil coleoids.

Belemnites range from the early Devonian to the end of the Cretaceous but are typically found in the Jurassic and Cretaceous. The State Fossil of Delaware, USA, is a Cretaceous belemnite, *Belemnitella americana*. Belemnites disappeared from the fossil record at the same time as ammonites and various other groups during the KT mass extinction; some belemnite-like fossils of younger age do occur but these seem to be more closely related to squids. In UK folklore belemnites are often referred to as 'thunderbolts' in reference to their bullet-like shape and the misguided belief that they were cast down from the skies during thunderstorms. The bullet-shaped belemnite 'guard' or 'rostrum' is a robust internal shell. Fractured examples show it to be a solid structure made of coarse, radiating crystals of calcite. The lack of an obvious analogue to the pointed guard among living animals provided a challenge to early naturalists who were unable to place belemnites in a major taxonomic group. The true affinities of belemnites only become clear when other components of the skeleton are preserved. The blunt end of the guard has a conical depression, called the alveolus, into which is inserted a chambered shell, straight and uncoiled but otherwise resembling the shell of living nautilus, that extends out beyond the depression. In contrast to the durable calcite guard, this phragmocone is fragile and made of

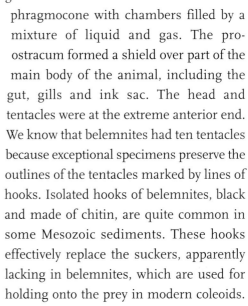

the relatively soluble mineral aragonite, thereby explaining why it is not often fossilised. Even less commonly found is a third component of the belemnite skeleton; the pro-ostracum, an aragonitic plate forming a prolongation on one side of the phragmocone.

The living belemnite animal would have strongly resembled a squid. The guard, enveloped by fleshy tissue, was located at the back (posterior) end of the elongate animal, occupying about one-third of its total length. At the centre of the animal was the phragmocone with chambers filled by a mixture of liquid and gas. The pro-ostracum formed a shield over part of the main body of the animal, including the gut, gills and ink sac. The head and tentacles were at the extreme anterior end. We know that belemnites had ten tentacles because exceptional specimens preserve the outlines of the tentacles marked by lines of hooks. Isolated hooks of belemnites, black and made of chitin, are quite common in some Mesozoic sediments. These hooks effectively replace the suckers, apparently lacking in belemnites, which are used for holding onto the prey in modern coleoids.

Belemnites are primitive among coleoids in that they used the chambers of the phragmocone to control buoyancy, much like the living nautilus and extinct ammonoids. More sophisticated chemical methods of buoyancy control have evolved in advanced swimming coleoids, involving ammonium replacement of denser metallic ions and chlorine replacement by sulphate ions in the soft tissues, as well as storage of low density oils in the liver. Swimming in belemnites was probably achieved to some extent by jet propulsion but it is also possible that the body covering the guard was

**Above** These two Jurassic belemnites – *Youngibelus tubularis* (left) and *Youngibelus levis* (right, 7.5 cm long) – have been interpreted as the male and female of a single species.

equipped with fins. Lateral grooves in some belemnite guards could indicate the attachment positions of such fins, which may have enhanced stability and manoeuvrability and perhaps generated the thrust for swimming at slow speeds.

Like ammonites, evidence has been found for the existence of differences between the sexes (sexual dimorphism) among belemnites. Two species of a Jurassic belemnite genus called *Youngibelus* that occur together in the same Jurassic rocks in Yorkshire have identical juvenile morphologies but differ substantially in their adult stage. While *Y. levis* is of conventional appearance, *Y. tubularis* has a tubular prolongation called an epirostrum extending beyond the pointed end of the guard. These two 'species' may not be true species but male and female individuals of one and the same species, although which is which remains unknown. The epirostrum could have counterbalanced the greater weight either of the body of a female brooding its young, or of an elongated arm used by the male in reproduction.

Sedimentary rocks occasionally contain belemnite guards in massive profusion. Explanations for these so-called 'belemnite battlefields' are numerous. Some are lag deposits formed by winnowing away of mud and silt, leaving behind the resilient guards of belemnites that belonged to many different generations and accumulated over a long period of time. Other concentrations could be 'graveyards' where moribund individuals died, possibly after reproducing. Yet others could represent concentrations of the waste products of animals that ate belemnites and either voided the indigestible guards after passage through their guts, or vomited them out through their mouths. We know that ichthyosaurs ate belemnites; masses of belemnite guards have been found within the body cavities of these dolphin-like reptiles where the stomach would have been located.

Belemnites are useful fossils in stratigraphy. Subtle differences in shape allow a large number of species to be recognised and, to the advantage of stratigraphers, individual species are often widely distributed, reflecting the free-swimming ecology of the living animals. However, barriers to species dispersal are evident.

**Above** 'Belemnite battlefield', a dense accumulation of belemnites covering a 22 cm wide slab of Jurassic shale from Yorkshire, England.

**Fig. 1 above** Living coral *Montastrea cavernosa*, photographed in the Florida Keys, showing the closely-packed polyps.

**Fig. 2 above right** Feeding tentacle crowns of some zooids in a living bryozoan colony.

**Fig. 3 right** Silicified colony of the tabulate coral *Syringopora* from the British Carboniferous. The tubular corallites are about 2 mm in diameter.

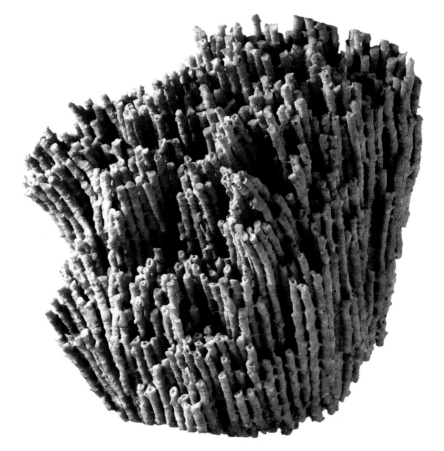

**Fig. 4 left** A small block of Ordovician shale from Estonia containing two colonies of the fenestrate bryozoan *Chasmatopora*.

**Fig. 5 below** Two living colonies of the mobile bryozoan *Selenaria* using their whisker-like setae to perform acrobatics.

**Fig. 6 opposite** Pipe-shaped living sponge photographed in Bonaire at a depth of 10 m.

**Fig. 7 above** The sphinctozoan sponge *Barroisia*, 3.8 cm wide, from the Cretaceous Faringdon Sponge Gravel of southern England, showing its modular construction.

**Fig. 8 right** The two 'tuning-fork' graptolites on this piece of Welsh Ordovician shale belong to the genus *Didymograptus* and have branches some 5 cm in length with sawtooth-like thecae.

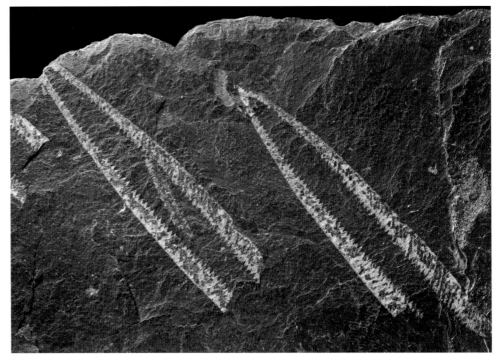

**Fig. 9 above** In folklore *Gryphaea* is known as Devil's Toenail; it is related to the living oyster. The strongly curved left valve dominates this lateral view of a Jurassic specimen, 7 cm long.

**Fig. 10 right** The gaping shell of the giant living bivalve *Tridacna* exposes to the light fleshy tissues that harbour symbiotic algae.

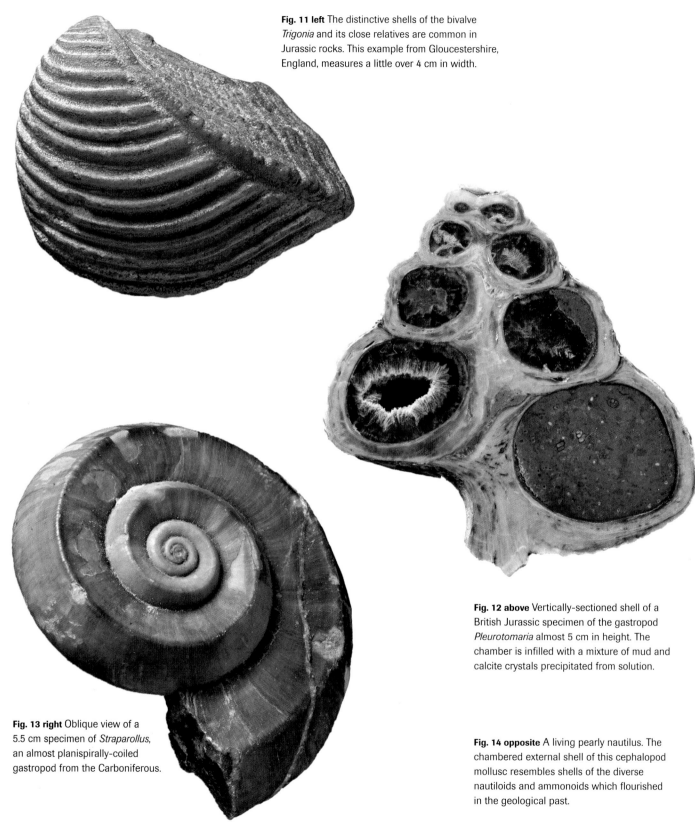

**Fig. 11 left** The distinctive shells of the bivalve *Trigonia* and its close relatives are common in Jurassic rocks. This example from Gloucestershire, England, measures a little over 4 cm in width.

**Fig. 12 above** Vertically-sectioned shell of a British Jurassic specimen of the gastropod *Pleurotomaria* almost 5 cm in height. The chamber is infilled with a mixture of mud and calcite crystals precipitated from solution.

**Fig. 13 right** Oblique view of a 5.5 cm specimen of *Straparollus*, an almost planispirally-coiled gastropod from the Carboniferous.

**Fig. 14 opposite** A living pearly nautilus. The chambered external shell of this cephalopod mollusc resembles shells of the diverse nautiloids and ammonoids which flourished in the geological past.

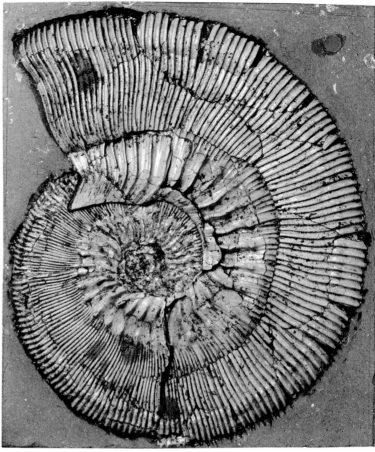

**Fig. 15 above left and fig. 16 above** Sexual dimorphism in the Jurassic ammonite *Lobokosmoceras*. The microconch possesses a spiny shell with a prominent lappet projecting from the aperture, whereas the macroconch has a plainer but larger shell 11 cm in diameter.

**Fig. 17 left** *Scaphites*, a Cretaceous ammonite, preserving the colourful nacre of the shell in this 7.6 cm wide specimen from South Dakota.

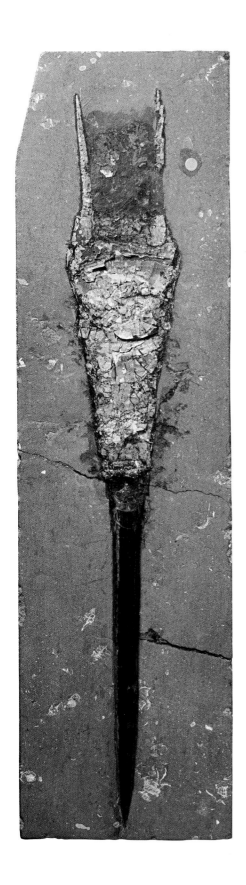

**Fig. 18 left** Measuring 22.5 cm in total length, this specimen of the Jurassic belemnite *Cylindroteuthis* preserves not only the bullet-shaped guard but also the crushed phragmocone and pro-ostracum.

**Fig. 19 right** Arms equipped with hooks and a black ink are visible in this 15 cm long, exceptionally preserved specimen of the coleoid *Acanthoteuthis* from the Jurassic of Wiltshire, England.

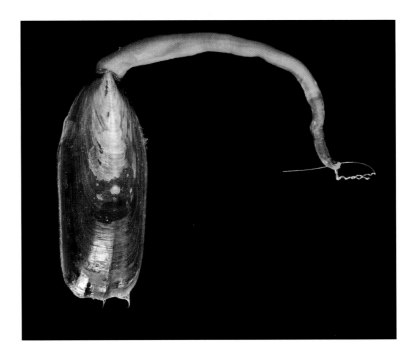

**Fig. 20 above** A living polyplacophoran mollusc, or chiton, showing the eight plates that make up the shell.

**Fig. 21 above right** Recent specimen of the brachiopod *Lingula* with a long pedicle emerging from the 5 cm long valves of the phosphatic shell.

**Fig. 22 right** The Christmas-tree worm *Spirobranchus* is a serpulid polychaete that employs a spiral crown of tentacles for capturing plankton.

**Fig. 23 right** Clusters of fossil serpulid worm tubes, like these *Rotularia* from the British Eocene, are common fossils in many Mesozoic and Cenozoic marine rocks.

**Fig. 24 below left** This example of the Silurian trilobite *Dalmanites*, 4.2 cm long, shows to perfection the complex dorsal exoskeleton of these extinct arthropods.

**Fig. 25 below right** The living horseshoe crab *Limulus* photographed off the coast of Florida.

**Fig. 26 above left** While amber is famous as a source of exquisite fossil insects, it also contains spiders such as this specimen of *Esuritor* entrapped in Baltic Amber of Eocene or Oligocene age.

**Fig. 27 above** Goose-necked barnacles showing the appendages used for feeding.

**Fig. 28 left** With a wingspan close to 7 cm, *Turanophlebia* is one of several dragonflies recorded from the Jurassic Solnhofen Limestone of Bavaria.

**Fig. 29 right** Irregular echinoid *Echinocardium cordatum* burrowing in sand. The flat-ended spines that propel this sea-urchin through its burrow are clearly visible.

**Fig. 30 far right** *Ova canalifera*, 7 cm long, from the Pliocene of France; apical view.

**Fig. 31 below** The long spines of the regular echinoid *Echinus acutus* from the northeast Atlantic provide a formidable defence in this surface-dwelling species which can grow up to 16 cm in diameter.

**Fig. 32 left** *Scyphocrinites elegans*, from the Upper Silurian or Lower Devonian of Erfoud, Morocco. View of a large slab (42 cm wide) with several individuals.

**Fig. 33 above** Two individuals of the living stalked crinoid (sea-lily) *Neocrinus decoras*, photographed from a submersible at a depth of 420 m in the Bahamas.

**Fig. 34 right** The free-living crinoid (feather star) *Antedon petasus* in which the ten pinnule-bearing arms may grow up to 10 cm long.

**Fig. 35 above** A dense population of the black brittle star *Ophiocomina nigra* photographed in the North Atlantic. Two echinoids are visible among the tangle of ophiuroids.

**Fig. 36 left** The asteroid (cushion star) *Asterina pectinifera* from an intertidal rockpool in Akkeshi Bay, northern Japan.

**Fig. 37 right** *Calliderma smithiae*, oral surface, 17 cm across the arms, from the Cretaceous Grey Chalk of Kent.

**Fig. 38 left** Reaching up to 50 cm in length, the holothurian (sea cucumber) *Stichopus tremulus* is here seen cohabiting a rocky platform with some regular echinoids.

**Fig. 39 below** *Deltoblastus*, 1.5 cm in diameter, from the Permian of Timor; oral, side and basal views (left to right).

Belemnites were evidently sensitive to water temperature; many groups, especially in the late Cretaceous, preferring colder waters and being absent in the tropics. Thus belemnites are generally less common in Tethyan deposits than at higher latitudes.

The oldest fossil octopus known is from the Jurassic of France, although an octopus-like fossil has recently been found in the Carboniferous of Illinois, USA. These shell-less coleoids are only found as body fossils in rare circumstances where soft tissues are preserved. However, the feeding activities of octopus may leave trace fossils in the shells of other molluscs. These take the form of tiny holes through which nerve toxin was injected to paralyse the unfortunate prey. Up to half of the dead shells in some populations of modern nautilus contain such holes, indicating that octopus are important predators of their cephalopod cousins.

### ACANTHOTEUTHIS: JURASSIC OF EUROPE AND ANTARCTICA

A coleoid related to belemnites, this genus (see colour fig. 19) is known particularly from some exceptionally preserved specimens in which the soft tissues were replaced by phosphatic minerals soon after burial. The body is up to 25 cm (10 inches) long and consists of eight arms with hooks, a small head and a sac-like tunic, within which is a conical phragmocone and short guard, with two lateral fins.

Spectacular examples of *Acanthoteuthis* (formerly called *Belemnotheutis*) were collected during the nineteenth century from the Jurassic Oxford Clay of Christian Malford in Wiltshire, England. Along with another coleoid *Mastigophora*, these often have the ink sac preserved.

### CYLINDROTEUTHIS: JURASSIC–CRETACEOUS OF EUROPE AND NORTH AMERICA

A belemnite with a typically large guard, up to 25 cm (10 inches) long, that is cylindrical, tapering gradually to a point. An anterior groove runs along one side of the guard. The depth of the alveolus may be up to a quarter of the length of the guard. Some examples preserve the broad, conical phragmocone almost as long as the guard (see colour fig. 18).

Guards of this belemnite are common in the Jurassic Oxford Clay of England.

### GONIOTEUTHIS: CRETACEOUS OF EUROPE

A belemnite with a moderately large guard, up to 9 cm (4 inches) long, ornamented by fine surface granules. The pointed end of the guard has a short but distinct knob-like extension called a mucro. Two lateral depressions and an anterior groove run along the guard. This belemnite has been used extensively as a zonal fossil in the late Cretaceous chalks of northern Europe. There are several species, each with a limited stratigraphical range, but these can be difficult to distinguish without making detailed measurements of such features as the length of the guard and depth of the alveolus.

**Above** Guard of the belemnite *Gonioteuthis*, 8 cm long, from the Cretaceous Chalk of Salisbury, England.

# Monoplacophorans

An extraordinary discovery in 1952 caused great excitement among biologists; an animal belonging to the molluscan class Monoplacophora, previously thought to have been extinct for some 350 million years, was found living in the Pacific Ocean off Mexico. Further living species of monoplacophorans have since been discovered, and the class is now known to have a worldwide distribution but with most species 'hidden from view' in deep water environments. Monoplacophorans have shells of aragonite and are univalved. In Recent species the shell is typically planispiral and limpet-like, but the fossil record also contains species with straight or spirally coiled shells. Unlike limpets and other gastropods, monoplacophorans do not undergo torsion. One to several pairs of pedal muscles clamp the soft body to the shell, leaving symmetrical muscle scars on the inside of the shell. Segmentation is more apparent in the soft parts of monoplacophorans than in any other class of shelled molluscs. This has been taken as evidence that monoplacophorans are primitive, which is consistent with their early geological appearance in the Cambrian. Although several hundred species of fossil monoplacophorans have been described, there is a huge gap in the fossil history of the group between the Devonian and Recent. The gap perhaps marks the retreat of monoplacophorans into deep water environments where they are less likely to enter the fossil record. This is due to the typically slow rates of sedimentation in the deep sea that do not favour fossilisation, as well as the more limited availability of rocks formed in the deep sea from which palaeontologists can collect fossils.

## Bellerophontids

This extinct group of molluscs with planispirally-coiled, bilaterally symmetrical shells ranges from the Cambrian to the Triassic. There is usually a notch at the shell aperture that becomes infilled in older parts of the shell to form a selenizone like that found in the pleurotomariid gastropods described above and suggesting a close affinity with gastropods. On the other hand, some bellerophontids show paired pedal muscle scars on the inside of the shell, as in monoplacophorans. This conflicting evidence has made the classification of bellerophontids controversial; some palaeontologists

believe that they are monoplacophorans, some that they are gastropods, and others that they consist of a mixture of monoplacophorans and gastropods.

## Polyplacophorans

At first glance resembling wood-lice (isopod crustaceans) because of their segmented upper surfaces, polyplacophorans are in fact a class of molluscs with shells consisting of eight plates (see colour fig. 20). The snail-like underside of the animal immediately betrays their true identity as molluscs rather than isopods. Like many snails, polyplacophorans are grazing animals, using the radula to rasp vegetation off the surfaces of rocks and shells in intertidal and shallow subtidal habitats. Unlike snails, however, they lack both tentacles and true eyes. Some 600 species of polypacophorans, or 'chitons' as they are commonly known, live today. Most are only a few centimetres long but the gumboat chiton – *Cryptochiton stelleri* – from the Pacific Ocean off western North America can grow to a massive 30 cm (12 inches) in length.

The combination of a shell that is made of aragonite and that disarticulates into its component plates after the soft parts of the animal have decayed, accounts for the rather poor fossil record of polyplacophorans. Fossils with the plates still articulated are very scarce. Instead, fossil polyplacophorans generally consist of isolated plates varying in shape from flat and semi-elliptical to arched and cusp-like. The oldest polyplacophorans date from the Cambrian. It has been claimed that the most primitive polyplacophoran species had seven rather than eight plates. This conclusion was reached by counting the number of different kinds of plates present in Cambrian assemblages of isolated polyplacophoran plates. Such a finding might be questioned as an error in the method of estimating plate number, were it not for the fact that during the development (ontogeny) of modern polyplacophorans seven plates are formed simultaneously and the eighth at a later time. During the evolution of polyplacophorans it appears that an eighth plate was for some unknown reason added to a seven-plated ancestor, the evolutionary (phylogenetic) increase in plate number being paralleled by an increase during the development (ontogeny) of individual polyplacophorans. The nineteenth-century German naturalist Ernst Haekel devised his 'Law of Recapitulation' for cases like this where ontogeny follows the same course as phylogeny. Recently, it has been suggested that a rare Cambrian–Carboniferous group of molluscs called the Multiplacophora with more than 17 plates were close relatives of the polyplacophorans.

# Rostroconchs

Not until 1972 were the rostroconchs recognised as a distinct class of molluscs; before this date they were usually included within the bivalves, although some examples were actually thought to be arthropods. Rostroconchs certainly resemble bivalves at first sight, but closer inspection shows that they lack the hinge found in true bivalves. Instead, the rostroconch shell is 'pseudobivalved', the two mirror image components being fused together. About 400 species of rostroconchs have been described. These range from the Cambrian to

**Above** Only 14 mm wide, this specimen of the rostroconch *Conocardium* from the Carboniferous of Belgium has a prominent rostrum (right).

Permian. They secreted shells of aragonite and hence are commonly preserved as moulds. Most rostroconchs are thought to have burrowed to shallow depths in soft sediments, but some may have lived on the surface of the sea bed. One or two apertures formed by gapes at the margins of the shell would have allowed the mollusc to extend its foot for burrowing and/or feeding. Some palaeontologists have interpreted rostroconchs as ancestral to bivalves. It is certainly easy enough to envisage the formation of a hinge line and a truly bivalved condition by progressive decalcification of a band at the dorsal edge of the calcareous shell of a rostroconch.

# Scaphopods

Sometimes known as tusk shells, scaphopods appeared rather late in the fossil record, the earliest unequivocal examples dating from the Carboniferous, almost 200 million years after the appearance of the other classes of molluscs (putting aside some very doubtful scaphopods from the Ordovician and Devonian). About 500 species of scaphopods live today, with a roughly equal number known only as fossils. All have gently curved, conical tube-like shells of aragonite that are open at both ends and range in length from a few millimetres to over 10 cm (4 inches). The surface of the shell is often plain, so that the shell resembles the stem of an old clay pipe. However, in some species the shell bears an ornament of ridges parallel to its long axis. Scaphopods burrow into sediments using the foot that protrudes from the opening at the wider end of the shell. About half to two-thirds of the shell is buried in sediment, the narrower end protruding above the sea bed. They feed on single-celled foraminifera and other small animals. Many species live in moderately deep water on the outer continental shelf and slope. An increase in scaphopod abundance is evident through the Mesozoic and Cenozoic. Locally, they occasionally occur in sufficient numbers to form shell beds. The position of scaphopods

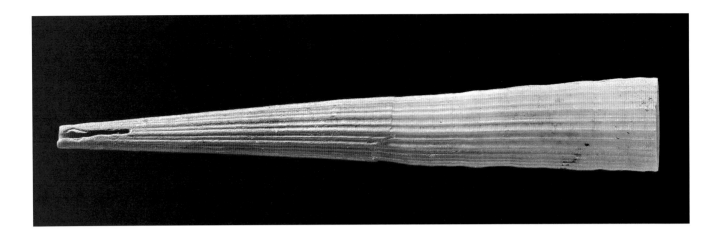

**Above** This tusk shell – scaphopod - from the Pliocene of Sicily measures 8.7 cm in length.

within the molluscs is unresolved; while some scientists believe that they are descended from rostroconchs, others place them closer to cephalopods.

## BRACHIOPODS

Brachiopods are a phylum that can be described with some justification as 'down on their luck'. Today they are uncommon and represented by fewer than 350 species in the seas of the world, a pale shadow of their former diversity and abundance in the geological past, especially during Palaeozoic times. About 4500 brachiopod genera have been described but more than 95 per cent of these are extinct. Few people are ever likely to have seen a brachiopod alive. Even divers and marine biologists seldom encounter brachiopods, as the greatest number of species live at depths between 100 and 200 m (330–660 feet). Yet for a good proportion of Phanerozoic time brachiopods were the dominant shelled animals living on the sea bed, and brachiopods are exceedingly common fossils in sedimentary rocks all around the world. But it is not only for reasons of sheer abundance that brachiopods have been extensively studied by palaeontologists. They are often useful in dating rocks, and as indicators of ancient environments. Furthermore, many palaeontologists have been intrigued by the vexed issue of whether the decline in brachiopods was in any way linked to the almost simultaneous rise in bivalve molluscs which they superficially resemble.

The surveyor and 'father of stratigraphy' William Smith (1769–1839) first knew of brachiopods by the vernacular name 'pundibs' used in some parts of southern England. A more usual common name for them is lamp-shells, alluding to the passing resemblance of certain brachiopods to Roman lamps. Brachiopods are suspension-feeding invertebrates, generally considered to have a kinship with bryozoans but to be even more closely related to the soft-bodied phoronid worms. In all species of brachiopods, the soft tissues of the animal are enclosed and protected by a mineralised shell consisting of two valves. The shell most often consists of calcite, but is occasionally made of organophosphate, and was probably aragonitic in one small extinct group. Calcitic brachiopods in particular have chemically stable shells that survive extremely well as fossils. Muscles are employed both to open and to close the shell, in contrast to bivalves where shell opening is achieved by means of an elastic ligament. The shell closing muscles are called adductors, and in most brachiopods the shell opening muscles are called diductors and pull against a fulcrum called the cardinal process.

The soft tissues of living brachiopods between the valves are surprisingly sparse, in fact, as much as 50 per cent of the organic mass of a brachiopod actually resides in the shell itself, either as protein sheaths wrapping around the minute crystals of calcium carbonate, or filling vertical canals called punctae. The most prominent organ is the lophophore used for feeding and also respiration. This occupies the mantle cavity between the two valves and is not protruded (unlike the lophophore of bryozoans). It is a bilobate structure, the lobes varying in shape according to species, from a simple crescent to a complex spiral which can coil in two or three dimensions. Long filaments (or tentacles) originate along the base of the lophophore. Beating of lateral cilia on the filaments drives water through the gaps between them, allowing planktonic food particles to be filtered off in the process. The particles are transferred by frontal cilia down the filaments to a brachial groove at the base of the lophophore, and thence to the mouth. Feeding can only occur when the valves of the shell gape enough to allow passage of plankton-laden water into the mantle cavity. Inhalent currents usually enter between the valves laterally, while exhalent

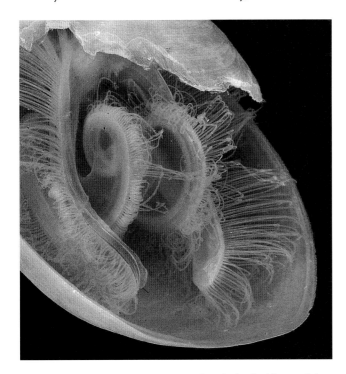

**Above** A recent brachiopod, *Mecandrevia* from Iceland, with one of the valves partly removed to expose the lophophore.

currents exit from the median area of the shell. In some brachiopods the lophophore is supported by a mineralised brachial skeleton or brachidium. The morphology of this brachidium is useful in distinguishing Mesozoic and Cenozoic fossil brachiopods with externally similar shell shapes.

Most brachiopods attach themselves permanently to a substrate using a fleshy structure called the pedicle, which is the only soft tissue permanently outside the shell in brachiopods. The pedicle emerges either from an opening in one of the valves, or from a gap between the two valves. The first condition is by far the most common and characterises a group of brachiopods formerly known as the Class Articulata, but now referred to as the Subphylum Rhynchonelliformea. The second condition occurs in most species belonging to a smaller group, once part of the Class Inarticulata, and now called the Subphylum Linguliformea. Some other inarticulate species with no pedicle are cemented to the substrate rather like oysters. These are placed in the Subphylum Craniiformea and differ also in having calcitic shells contrasting with the organophosphatic shells of the Linguliformea. The old terms articulate and inarticulate refer to the presence or absence respectively of articulations between the two valves of the shell.

Cambrian brachiopods are predominantly linguliforms. The early evolutionary radiation of this group with organophosphatic shells occurred at a time when calcium phosphate seems to have been favoured for skeleton construction by many marine animals. Specimens of the linguliform *Micromitra* from the Burgess Shale of British Columbia (Canada) are remarkable in preserving setae (hair-like structures protruding between the two valves and equalling or exceeding the length of the shell itself). Setae occur in all living species of brachiopods but are not mineralised and therefore seldom preserved in fossils. A sensory function seems most likely for the setae of *Micromitra*, perhaps detecting the approach of predators or particles that might damage the lophophore and eliciting protective closure of the shell. Although linguliformeans have been greatly outnumbered by other brachiopods since their zenith in the Cambrian, they have persisted through to the present day. The best known extant linguliformean is *Lingula* itself (see colour fig. 21), intriguing not only for its fame as a living fossil that has at first sight changed little since the Ordovician, but also because of its abnormal ecology. Unlike other brachiopods, *Lingula* lives in vertical burrows excavated into fine sands in brackish or intertidal environments, using its setae to form three siphon-like structures at the top of the burrow for the passage of inhalant and exhalant feeding currents. The very substantial pedicle of *Lingula*, which can be used to haul the shell deeper into its burrow, is a culinary delicacy in some parts of Asia, making this the only brachiopod to be eaten by humans. Fossil *Lingula* and related genera usually have thin valves, almost circular to tongue-shaped, and tending to be shiny and dark in colour because of their phosphatic composition.

## BRACHIOPOD CLASSIFICATION AND GEOLOGICAL RANGES

Note that only a small selection of the large number of brachiopod orders are included here.

Subphylum Linguliformea (Cambrian–Recent)
Subphylum Craniiformea (Cambrian–Recent)
Subphylum Rhynchonelliformea
    Order Strophomenida (Ordovician–Triassic)
    Order Productida (Devonian–Triassic)
    Order Orthida (Cambrian–Permian)
    Order Pentamerida (Cambrian–Devonian)
    Order Rhynchonellida (Ordovician–Recent)
    Order Spiriferida (Ordovician–Permian)
    Order Terebratulida (Devonian–Recent)
    Order Thecideidina (Triassic–Recent)

The hard surfaces furnished by dead shells and echinoid tests in the late Cretaceous Chalk sea of northern Europe were often utilised as attachment substrates by the craniiformean brachiopod *Ancistrocrania*. One of the valves (ventral, formerly called the pedicle valve) was cemented over its entire basal area to the substrate. The other valve (dorsal, formerly the brachial valve) was attached to the first by muscles when the brachiopod was alive and invariably became separated after death. Therefore, ventral valves are best located by looking over the surfaces of shelly substrates, whereas dorsal valves can be found by sifting sediment samples. The dorsal valve resembles a limpet but, of course, is unrelated to these gastropod molluscs. Measuring about 1 cm (0.4 inches) in diameter, the ventral valve has thickened, rampart-like edges and an interior surface with deep scars where the muscles were anchored to the shell.

While both linguliformean and craniiformean brachiopods are locally abundant, the bulk of the fossil record of brachiopods comprises rhynchonelliforms, the old articulate brachiopods. These are also the brachiopods most likely to be confused with bivalve molluscs. Contrasts in symmetry, however, usually enable them to be told apart: in brachiopods the two valves are dissimilar but each valve is bilaterally symmetrical about its own mid line; in bivalve molluscs, the two valves are normally mirror images but each valve is not bilaterally symmetrical about its own mid line. The valves of a brachiopod are designated dorsal and ventral, those of a bivalve mollusc left and right. Unfortunately, there are some exceptions to these rules of symmetry and other characteristics must be considered as well. For example, most brachiopods have a hole (pedicle foramen) in the ventral valve. There is no immediate analogue to this structure among bivalve molluscs. Shells covered by a regular pattern of minute holes (punctae) characterise some brachiopods, especially in the Mesozoic and Cenozoic, but never occur in bivalves. The presence of these punctae can be

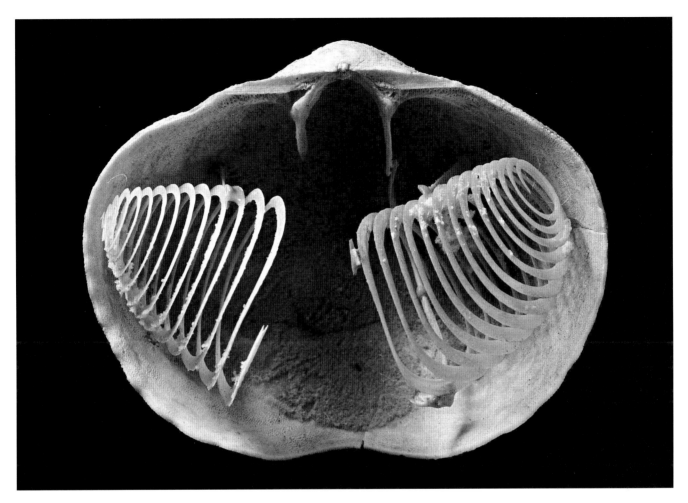

**Above** Remarkable preservation of the delicate brachidium which supported the lophophore in a French Jurassic specimen of the spiriferide brachiopod *Spiriferina*.

especially helpful in determining the identity of shell fragments. Moulds of brachiopods in unweathered limestones are rare because their stable calcitic shells seldom dissolve. In contrast, preservation of bivalves as moulds is common because many species have aragonitic shells prone to dissolution.

Palaeozoic rhynchonelliform brachiopods come in a variety of shapes and sizes, far exceeding the range seen in their descendants that survived the catastrophic extinction at the end of the Permian period. Some have conventional biconvex shells in which both the dorsal and ventral valves are outwardly bowed. Others, however, have concavo-convex shells. Here the ventral valve is convex and larger than the dorsal valve that sits inside it and has a hollow, saucer-like shape. The opening for the pedicle in concavo-convex brachiopods is commonly tiny or sealed, and these brachiopods evidently lived freely on muddy yet firm sea beds. Most probably the larger ventral valve rested on the sediment surface. The curvature of concavo-convex shells often includes an inflection beyond which is a part of the shell called the trail. In genera such as *Leptaena* from the Silurian, the existence of a trail would have ensured that the shell gape was elevated above the level of the sea bed and sediment less likely to become entrained in the feeding currents. Some of the largest brachiopods are concavo-convex species from the Carboniferous and Permian belonging

CHAPTER THREE

to a group called the productines. These may have very thick shells, particularly in the early-formed beak region of the ventral valve. This extra weight would have been important in providing stability to impede overturning, while the large surface area of the ventral valve acted as a snowshoe in spreading the weight and preventing the animal from sinking too deeply into the sediment. Additional support and anchorage is obtained in some species from long hollow spines that may extend for a considerable distance beyond the margins of the valves.

Orthides are an ancient group of rhynchonelliform brachiopods particularly common in some early Palaeozoic rocks. Their shells may form thin shell beds or coquinas in shallow water marine sediments. In outline shape, orthides are commonly almost semicircular, the two valves being joined along a straight hinge line. Ribs may radiate outwards from the centre of the hinge line.

Another group of rhynchonelliforms, the spiriferides, have shells with lateral prolongations forming tapering 'wings'. A spiriferide found in Devonian slates of Cornwall, England is known locally as the 'Delabole Butterfly' on account of the wings that are very long. Each wing of the spiriferide shell contained an arm of the lophophore coiled into the shape of a helicospiral. This is known because the mineralised support of the lophophore – the brachidium – is occasionally preserved intact within the shell. Spiriferide brachiopods in the Devonian were favoured substrates for smaller animals. In some populations, shells are profusely encrusted by bryozoans, corals and worm-like tubular fossils called cornulitids. The distributions and growth directions of these symbionts have been used by palaeontologists as clues to the life orientations and feeding currents of the host brachiopods. For example, preferential growth of cornulitids towards the lateral edges of shells has been taken to indicate that inhalent feeding currents were located here and taken advantage of by the symbiotic worms. In spiriferides, as in many other brachiopods,

the junction between the valves, known as the commissure, contains a median fold forming a sulcus. This is thought to assist in separating inhalent from exhalent feeding currents. Inhalent currents entered at the sides of the shell, as predicted in spiriferides by the growth of symbiotic worms, and exhalent currents departed in the position of the sulcus.

The most bizarre brachiopods are two specialised groups that inhabited reefs during the Permian. Both became cemented to the reef surface. The lyttoniidines have irregular shells, the dorsal valve being greatly reduced and mainly comprising a highly lobate plate that supported the lophophore during life. The second group, the richthofenidids, (e.g. *Cyclacantharia*), resemble solitary corals, with a large conical ventral valve supporting a small lid-like dorsal valve. A theory that the dorsal valve of richthofenidids was flapped by muscular action to flush water into and out of the mantle cavity has fallen out of favour. Normal feeding using lophophore generated currents seems more likely to have occurred, although it is possible that these tropical reef-dwelling brachiopods harboured symbiotic algae like modern reef corals.

Brachiopods were common in tropical regions of the Palaeozoic. With the exception of one group, however, they are rare in the tropics today. Most modern brachiopods live in temperate or polar waters, and they are one of few marine phyla not to show an increase in diversity from pole to equator. The one group of brachiopods characteristic of the tropics today are the thecideidines. Ranging back to the Triassic, these tiny brachiopods are typically found cemented to the undersides of corals or rubble in coral reef environments. In such hidden (cryptic) habitats they are less likely to be found by predators. Brachiopods also discourage predators by manufacturing poisonous chemicals. Legend has it that a research student anxious to confirm the unpalatability of a living species ate a small amount on brachiopod flesh and became so unwell that he needed hospital treatment.

120

Thecideidines are a minor group compared to the two dominant groups of the post-Paleozoic, the rhynchonellides and terebratulides. Rhynchonellides generally have coarsely-ribbed shells resembling cockles (bivalve molluscs), whereas terebratulides are usually smooth-shelled. Shell ribbing is thought to have a strengthening function: in much the same way as corrugated aluminium used in roofing is stronger than a flat sheet of aluminium, so a ribbed shell is stronger than an unribbed shell. The ribs also have the effect of producing a zig-zag commissure between the two valves, with the result that a relatively small opening of the shell gives a large area for feeding currents to enter and exit yet does not allow large, unwanted

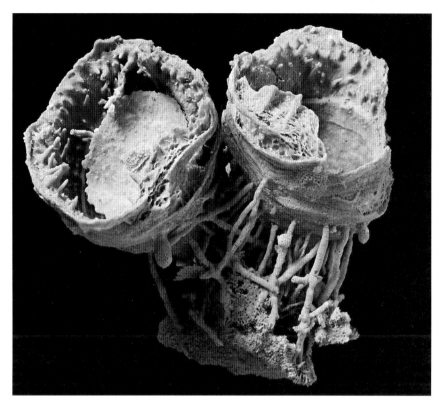

**Above** Two individuals of the peculiar Permian brachiopod *Cyclacantharia*, each about 2 cm in diameter, showing the solitary coral-like shape and long supporting spines.

particles to pass into the mantle cavity. One of the few brachiopods to break the rule of valve symmetry is a Jurassic rhynchonellide called *Torquirhynchia inconstans*. The somewhat rotund shells of this species have a single median deflection of the commissure such that on one side of the brachiopod the dorsal valve is deeper and on the other side the ventral valve is deeper. Left- and right-handed examples of *Torquirhynchia inconstans* can be found. The reason for this asymmetry is unknown.

The most common group of living brachiopods are the terebratulides. In terms of abundance, however, terebratulides enjoyed their heyday in the Mesozoic. Shallow marine carbonates of Jurassic and Cretaceous age frequently contain terebratulides, sometimes clustered together in 'nests'. During life, many of the individuals in a nest would have attached to others using their pedicles. Although the pedicles themselves do not fossilise, a trace fossil boring called *Podichnus* can be left in the substrate where the pedicle was once attached. The circular patch of minute holes characteristic of *Podichnus*

mark the anchorage points of individual threads of the pedicle. Brachiopod shells with surfaces covered by *Podichnus* can be found and indicate that these animals served as substrates for others in the nest.

Terebratulides are seldom found as fossils with the two valves disarticulated. This is because the teeth and sockets along the hinge line between the valves keep them joined together after the animal has died. Indeed, the valves of the shell often begin to disintegrate before disarticulation occurs. Proteins that wrap around and bind the fibrous calcite crystals of the shell during life begin to decay after death, causing the fibres to fall apart.

As mentioned at the beginning of this section on brachiopods, there is debate about whether the decline of brachiopods is in any way linked to the rise of bivalve molluscs. One extreme interpretation sees the two groups as 'ships that pass in the night', each effectively unaware of the others existence. Straightforward competition between the two groups may not explain the pattern completely. For one thing, many of the bivalves that diversified at the time brachiopods were declining were species that burrowed into sediment and were not in direct competition with the surface-dwelling brachiopods. The end Permian extinction event drastically reshaped many ecosystems and probably played a major role in the switchover from brachiopod to bivalve domination. Brachiopods were particularly hard hit by this extinction and were slow to recover. By the time they did so, bivalves had evolved to occupy many of the habitats previously monopolised by brachiopods.

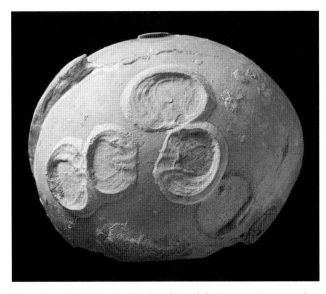

**Above** Several individuals of the brachiopod *Ancistrocrania* cemented to an echinoid from the Cretaceous Chalk of Kent, England, which is 5 cm in width.

## *ANCISTROCRANIA*: CRETACEOUS OF EUROPE

In this brachiopod the ventral valve is cemented to a hard surface over much of its outer surface. The shell is calcareous, about 1 cm (0.4 inches) in diameter, each valve being roughly circular in plan view with one straighter edge corresponding to the hinge line along which the valves are hinged during life (they are generally found separated in fossils). Internally, both valves have paired circular muscle scars and prominent 'vascular' markings. In addition, a pair of processes is visible on the inner surface of the dorsal valve.

Individuals of this brachiopod and related craniacean genera encrust hard surfaces in the sea, including rocks and shells. Specimens from the Cretaceous Chalk of northern Europe are frequently to be found on tests of echinoids, especially *Echinocorys*, where they were members of small-scale communities that developed on these hard surfaces between the death of the echinoid and the final burial of the test.

## *BEECHERIA*: CARBONIFEROUS–PERMIAN, WORLDWIDE

The shell of *Beecheria* has an elongate elliptical outline. Both valves are convex, the ventral valve somewhat more so than the dorsal valve. There is no shell ornamentation but a broad, subdued fold and depression (sulcus) may

**Above** Around the lower margin of this Irish Carboniferous shell of *Beecheria*, 1.9 cm wide, can be seen faint traces of an original pigmentation consisting of narrow stripes.

be present. A pedicle opening occurs at the apex of the umbo on the ventral valve.

This terebratulide brachiopod was tethered by its pedicle. Some examples from the Carboniferous preserve traces of an original shell pigmentation in the form of dark radial stripes around the shell margin.

**Above** *Clitambonites* from the Estonian Ordovician, showing the dorsal valve exterior and interior (upper figs.), and the ventral valve interior, the latter shell measuring 2.9 cm in width.

## *CLITAMBONITES*: ORDOVICIAN OF EUROPE AND ASIA

Shells have a rounded quadrangular outline, widest at or close to the hinge line. The ventral valve is markedly convex, the dorsal valve less so or even flat. They are ornamented externally by fine ribs and frilly growth lines. A large triangular 'interarea' is present where the ventral valve overlaps the dorsal valve, containing a deltidial plate beneath the small apical pedicle opening. Internally, teeth and sockets are well developed and the dorsal valve interior has a low median septum and small cardinal process.

*Clitambonites* is a strophomenide rhynchonelliform characterised by having a large deltidium, a calcareous plate that restricts the opening for passage of the pedicle in the ventral valve.

## *GIGANTOPRODUCTUS*: CARBONIFEROUS, WORLDWIDE

*Gigantoproductus* is a large, thick-shelled brachiopod with a convex ventral valve and concave dorsal valve. The shell is broad, the prolonged lateral edges giving it a wing-like appearance. An irregular ornament of costae covers the outer surface. Internally, the ventral valve has a twin-lobed impression apparently marking the site where the lophophore was fused to the mantle tissue lining the inside of the shell.

This productide genus is among the largest of all brachiopods, shells reaching 30 cm (12 inches) in width. Given its huge and thick shell, *Gigantoproductus* has a

**Above** A 16 cm wide specimen of the huge Carboniferous brachiopod *Gigantoproductus* showing the coarse ribbing on the ventral valve.

surprisingly narrow space between the valves to accommodate the lophophore and other soft tissues of the body. Any shell-breaking predators attacking *Gigantoproductus* would have expended a lot of energy for little return.

### *KIRKIDIUM*: SILURIAN OF EUROPE AND NORTH AMERICA

The ovoidal shells of *Kirkidium* reach up to 10 cm (4 inches) in length and have a large ventral valve with an incurved umbo that overhangs the smaller dorsal valve. Strong ribs are present on both valves. The interior of the ventral valve has a long median septum, that of the

**Above** Dorsal and lateral views of an 8.6 cm high specimen of *Kirkidium*, a pentameride brachiopod from the Silurian of the Welsh Borders.

dorsal valve two shorter, diverging septa; specimens often break along these septa during collection.

Found in Silurian limestones, the large pentameride rhynchonelliform *Kirkidium* was sometimes present in sufficient densities to have built reef-like biohermal structures.

### *LINGULA*: ORDOVICIAN–RECENT, WORLDWIDE

The two valves of the *Lingula* shell are almost the same shape and size except that the ventral valve is slightly longer than the dorsal valve. Both valves are gently convex and plectrum- or spade-shaped in outline (see colour fig. 21). A long, fleshy pedicle emerges at the apex of the shell between the valves in living specimens. Made of calcium phosphate, the shell in fossil specimens is typically shiny, brown or black in colour, and is ornamented by very fine growth lines. Internal teeth are lacking.

This unusual linguliform brachiopod lives in burrows. Hair-like setae extend from the anterior opening of living individuals, separating a median region of inhalent flow of water carrying food particles, from two lateral regions of exhalent flow of water taking away filtered water. Whereas *Lingula* tends to be an inhabitant of intertidal zones today, some fossil species lived in deeper waters.

### *MUCROSPIRIFER*: DEVONIAN, WORLDWIDE

This butterfly-shaped brachiopod is significantly broader than high, having its maximum width along the almost straight hinge line. Both valves are convex and carry strong ribs. Growth lines on the shell surface have the form of frills. A pronounced median fold in the dorsal valve matches a depression (sulcus) in the ventral valve. The beak is small.

Enormous numbers of this spiriferide rhynchonelliform occur at certain Devonian sites in North America; perfectly preserved and fully articulated shells can be collected by the bucket-load from some clays in Ohio and Ontario.

## *RAFINESQUINA*: ORDOVICIAN, WORLDWIDE

The shell of *Rafinesquina* is almost semicircular in outline, with a straight hinge line, and usually 3 or 4 cm (1–2 inches) in width. Externally it bears an ornament of fine ribs and growth lines. The ventral valve is convex and the dorsal valve concave and fitting within it, leaving a narrow cavity for the soft parts of the brachiopod. Internally, the ventral valve has teeth and prominent muscle attachment scars, whereas the dorsal valve has a large cardinal process for anchorage of the muscles that opened the shell.

This genus belongs to a group of rhynchonelliform brachiopods (strophomenides) that were very numerous and diverse during the Palaeozoic. Although juveniles may have been attached by a pedicle, adults rested freely on lime-mud sediments. There has been debate among palaeontologists about which valve was uppermost, the convex ventral valve or the concave dorsal valve, the latter being favoured by the majority.

**Above** *Mucrospirifer* showing, in dorsal (upper fig.) and ventral (lower fig.) views, the characteristically wing-like shell. This specimen from the Devonian of Ohio is 3.5 cm wide.

**Below** Two Ordovician specimens of *Rafinesquina*, about 4.5 cm wide, showing the exterior (left) and interior (right) of the dorsal valve.

**Above** The pedicle opening is very clear in this British Pliocene example of *Terebratula* measuring 6.6 cm in height.

## *TEREBRATULA*: MIOCENE–PLIOCENE OF EUROPE

The shell is of moderate to large size, roughly oval in outline, and both valves are convex. An ornament of growth lines is present but there are no ribs. The valves are thick, especially around the umbo, and the shell fabric contains punctae (pores). A large pedicle opening is present in the ventral valve. Internally, the ventral valve has hinge teeth and a collar around the pedicle opening. The interior of the dorsal valve has a rounded cardinal process and a short median septum.

The name *Terebratula* was formerly applied very broadly for numerous brachiopods of Mesozoic and Cenozoic age that are now accommodated in other terebratulide genera. In its modern sense *Terebratula* is

confined to the original, 'type' species from the Pliocene of Italy, plus some related species of similar age from elsewhere in Europe.

## *TORQUIRHYNCHIA*: JURASSIC–CRETACEOUS, WORLDWIDE

The shell of this rhynchonelliform brachiopod is of medium size and often quite globose. Both valves are convex, the dorsal valve somewhat more so than the ventral valve. In outline the shell is triangular, with its maximum width approximately midway between the

**Above** While the Jurassic *Torquirhynchia* is fairly typical of the Order Rhynchonellida in dorsal view (upper fig.), the bilateral asymmetry of the zig-zag commissure (lower fig.) is an unusual feature.

posterior and anterior extremities. Strong, sharp ribs radiate from the umbos of each valve, giving a zig-zag plication where the valves meet along the anterior commissure. Viewed from the anterior, the characteristic asymmetry of *Torquirhynchia* becomes evident, either the left or right side of the commissure being higher than the other side. The beak of the ventral valve is small, incurved, and has a tiny pedicle opening.

### *ZEILLERIA*: TRIASSIC–JURASSIC, WORLDWIDE

Shell shape in *Zeilleria* is variable but most species have lobate shells, some with four lobes giving the fossil a leaf-like appearance. Both valves are moderately convex, the ventral valve having a highly incurved beak with a pedicle opening. The shell surface is relatively smooth, although faint growth lines may be developed. Openings of the punctae that pass through the thickness of the shell as narrow canals are particularly evident when shells are examined using a microscope. Inside the dorsal valve are crural plates, a long loop that supported the lophophore, and a median septum.

In common with most other post-Palaeozoic terebratulides, secure identification of *Zeilleria* necessitates knowledge of internal structures. Because valves are generally articulated and the interior is filled with hardened sediment and/or calcite cement, these structures can only be revealed by the laborious preparation of serial sections at regular intervals through the fossil, a process which destroys the fossil itself.

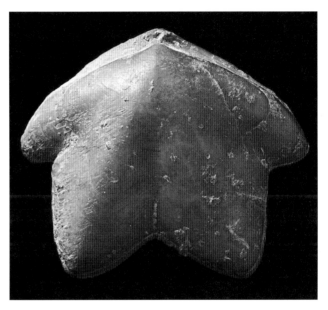

**Above** Leaf-shaped shell of a French Jurassic species of the brachiopod *Zeilleria* measuring 3.8 cm wide.

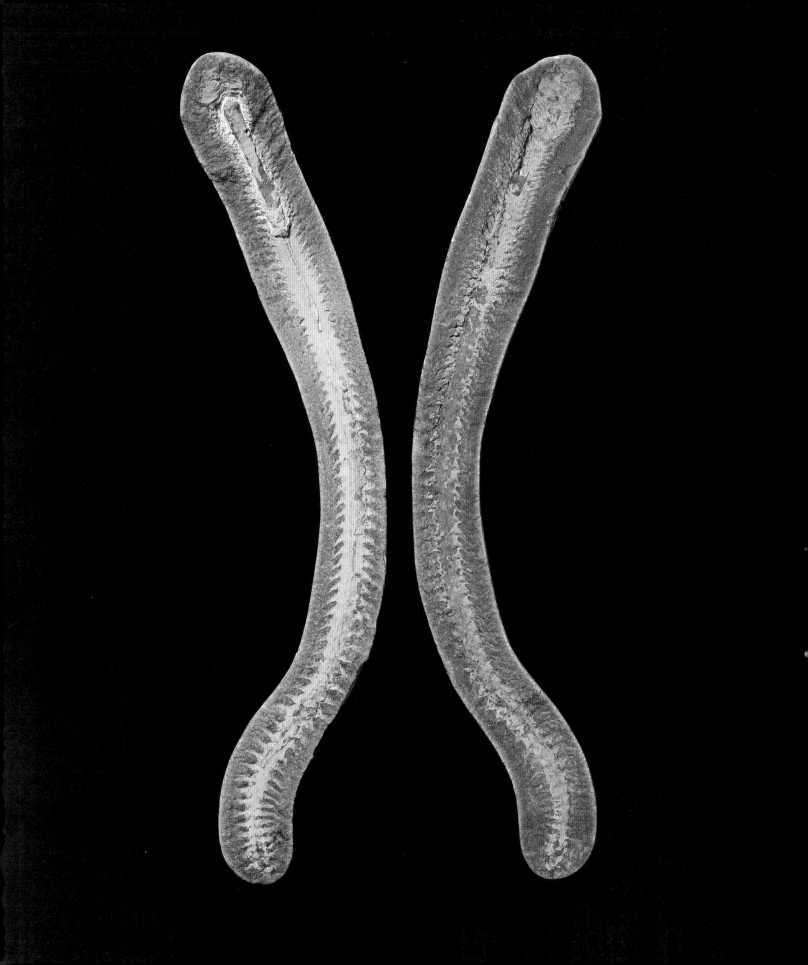

# 4

# Worms and tubes

TODAY THE ANNELIDA rank among the most abundant of all invertebrate phyla. Annelids are just one of several types of 'worm'; other worms belong to phyla such as the Nematoda, Nemertina, Platyhelminthes, Sipunculida, Onychophora, Phoronida and Priapulida. Because of their relatively minor contribution to the fossil record, at least as body fossils, it is not unusual for worms to be ignored in books about invertebrate fossils. However, there are some fossil worms with mineralised hard parts that are quite abundant as fossils, while rarer instances of soft-bodied fossil worms provide us with tantalising glimpses of the importance of these animals in the geological past. This short chapter also ventures beyond fossil worms of known affinities to consider three enigmatic groups of tubular calcareous fossils – tentaculitids, cornulitids and hyoliths – that might have been made by animals belonging to one of the worm phyla still living today.

## ANNELIDS

The Phylum Annelida contains about 15,000 living species, with different species inhabiting terrestrial, freshwater and marine environments. There are three subdivisions of annelids: polychaetes, oligochaetes and hirudineans. Polychaetes (bristleworms), the largest group, mostly live in the sea and are by far the best represented annelids in the fossil record. The name polychaete refers to the numerous (poly-) bristle-like chaetae on the body. By contrast, oligochaetes have fewer chaetae. The earthworm is a typical oligochaete terrestrial worm. Hirudineans (leeches) usually inhabit freshwater environments, many being blood-sucking external parasites of fish. There are no undisputed fossil examples either of oligochaetes or hirudineans.

Modern polychaetes encompass a wide range of lifestyles. They include carnivores, herbivores, omnivores and detritus-, suspension- and deposit-feeders; some are highly mobile, others sedentary or totally sessile. Predatory species that must go in search of their prey have to be mobile and are variously capable of swimming, crawling and burrowing. Suspension feeding polychaetes usually inhabit tubes of their own making and are sessile, while deposit feeders burrow into sand or mud.

A few Cambrian body fossils of polychaetes (and probable polychaetes) have been described, mostly from the Burgess Shale of Canada. These may well be the oldest fossil annelids, although it should be noted that some flattened, segmented fossils (*Dickinsonia* and

**Left** Astonishing preservation of a eunicid worm from the Pleistocene of Greenland. The worm's body formed the nucleus for the growth of a 20 cm long sinuous concretion, split into two after collection to reveal the fossil inside.

*Spriggina*) from the late Precambrian Ediacaran biota of South Australia have in the past been interpreted as worms. The Burgess Shale polychaetes *Burgessochaeta, Canadia, Insolicorypha, Peronochaeta* and *Stephanocolex* are clearly segmented, with most of the segments bearing brush-like clusters of chaetae along each side of the elongate body. Another famous Burgess Shale fossil called *Wiwaxia* has a flattened oval-shaped body adorned with plates and spines. It has been variously classified as a polychaete or a relative of the molluscs.

The fossil record of completely preserved soft-bodied polychaetes becomes very sparse after the Cambrian, perhaps a reflection of the increasing numbers of scavenging animals able to feed on their corpses and burrowing animals churning-up the sediment. However, two other kinds of polychaete body fossils do begin to appear in the fossil record. The first are polychaete jaws known to palaeontologists as scolecodonts. At least seven living families of polychaetes have jaws with small teeth that would be categorised as scolecodonts if fossilised. These teeth are made of chitin or scleroprotein and occur in one to several pairs along the proboscis of the worms. Sometimes the teeth contain a hollow cavity through which venom can be transferred from a poison gland for injection into the prey. Scolecodonts matching the jaws of the extant order Eunicida first appear in the early Ordovician, increase in diversity through the Ordovician and into the later Palaeozoic, but suffer significant extinctions at the end of the Permian before recovering during the Mesozoic. Like many modern eunicids, the scolecodont-producing fossil species were predators, some possibly sheltering in burrows in mud or crevices in harder rocks and emerging to feed, as in the extant genus *Marphysa*.

Other polychaete body fossils are the calcareous tubes constructed by serpulid worms. These are common in many sedimentary rocks formed in shallow water marine environments. Recent serpulids number some 500 species. They are known as fanworms on account of the fan-like crown of radioles (tentacles) that they use for

**Above** Scanning electron micrograph of the scolecodont *Oenites*, the 2 mm long jaw of an annelid worm from the British Silurian.

feeding on plankton. The typically brightly-coloured radiolar crown protrudes out of the tube for feeding but can be withdrawn back into the tube if menaced by predators or, in the case of worms living in the intertidal zone, when the tide recedes and desiccation threatens. In some species, an operculum, which can be calcified, plugs the tube when the radiolar crown is retracted. By the standards of polychaete worms, serpulids have relatively short bodies typically several times shorter than the length of the tubes they occupy. In addition to the radiolar crown, there is a thorax and a segmented abdomen. Serpulid worm tubes grow by the worm applying a rapidly solidifying paste of calcium carbonate and mucropolysaccharide, secreted from a pair of glands, to the open aperture of the tube.

Although serpulid opercula can be fossilised, it is their tubes that are most likely to be encountered by fossil collectors. Consisting of either calcite, or occasionally the less stable mineral aragonite, serpulid tubes are generally firmly cemented to a hard surface, such as a shell or rock, but can also be attached to seaweeds. Dense clusters of serpulids are common. Tubes vary in shape from tight spirals, as in the small spirorbids, to loosely coiled, meandering or straight cylinders (see p. 67). Some tubes have a rounded cross-section but others develop one or more ridges running along their lengths. Tube diameter may increase appreciably or hardly at all with growth, and in some species the tube begins to grow upright from the substrate after an initial encrusting phase. Serpulid tubes can reach over 10 cm (4 inches) in length. Among the largest living serpulids is the 'Christmas-tree worm' *Spirobranchus* (see colour fig. 22) which grows embedded in corals and can live for up to 40 years.

There are numerous records of supposed serpulid tubes in the Palaeozoic. However, these may have been made by other invertebrates as the microstructure of their tubes often shows significant differences from that of true serpulids. Furthermore, some of the Palaeozoic fossils assigned to the extant serpulid genus *Spirorbis* are found in sedimentary rocks deposited in freshwater environments, whereas true *Spirorbis* is an exclusively marine worm. Indeed, there is only one recent serpulid species that inhabits freshwater, the cave-dwelling *Marifugia cavatica*. By the Triassic, however, *bona-fide* serpulid tubes do occur in the fossil record, and they are found in ever increasing numbers through to the present day. Serpulids are an almost ubiquitous component of Jurassic and Cretaceous communities that encrusted living or dead brachiopod and mollusc shells. Often the tubes grew towards the shell gape, suggesting that the worm was taking advantage of the feeding currents generated by a living host animal. Elsewhere, serpulid tubes on inclined substrates can be found to have grown up-slope. This was probably an attempt to reach the highest point on the substrate, thereby escaping the turbid waters and danger of burial caused by deposition of muddy sediment on the sea bed

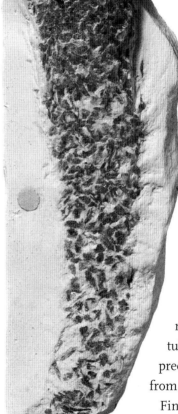

**Above** Tube constructed by a Cretaceous 'worm' which gathered and stuck together plant debris. The fossil is 7.5 cm long.

around. Reefs up to 2 m (6.5 feet) in diameter are constructed by aggregations of the serpulid *Rotularia* in the Eocene (see colour fig. 23). Unfortunately, matching fossil serpulid tubes with living genera can be difficult because zoologists identify serpulids mainly on the basis of their soft parts and pay little attention to the calcareous tubes.

A great deal of interest has focused on communities, first discovered in 1977, living around hot water vents in the deep sea along mid-oceanic ridges. One of the most characteristic animals present in these hydrothermal communities in the Pacific Ocean is the giant worm *Riftia* which belongs to a group of annelids called the vestimentiferans. This worm lives in a tube made of chitin, protruding its gills from the open end. Adults of *Riftia* lack a gut and rely on symbiotic bacteria for their nutrition. Investigations of ancient vent communities have revealed the presence of possible vestimentiferan tubes as far back as the Devonian and possibly the Silurian. These tubes can be preserved as external moulds resulting from overgrowth of the tubes by minerals, especially iron pyrites, precipitated when the hot fluids emerging from the vents met the cold sea water.

Finally, mention should be made of the diverse fossil 'worm' tubes made of foreign particles – sand grains, shell fragments, pieces of plant debris, etc. – collected by the tube-building animal and stuck together. Some of these agglutinated worm tubes are very similar to tubes made by living polychaetes, such as *Sabellaria*, but there is usually insufficient information to be categorical about the identity of the tube builder.

## NEMATODA

The Nematoda (roundworms) is a phylum of simple worms with tapering ends that are extremely abundant in the mud on the floor of the deep sea and also occur as internal parasites of animals and plants. Their fossil record is scant, although fossils of nematodes have been found preserved in Cenozoic amber, including examples that were parasites of ants some 40 million years ago.

## ONYCHOPHORA

Although represented today by fewer than 100 species, the Onychophora or 'velvet worms' may have been a much richer phylum in the geological past. These soft-bodied worms have between 14 and 43 pairs of stubby walking legs and at first sight resemble large caterpillars. Today they inhabit damp terrestrial environments but Cambrian examples lived in the sea: two genera of onychophorans occur in the Burgess Shale, a Cambrian marine *Lagerstätten*, and other Cambrian fossil onychophoran are known from marine deposits in China and Greenland. One of the Burgess Shale onychophorans is the infamous *Hallucigenia*, named because of its dream-like appearance, with the animal apparently supporting its barrel-shaped body above the sea bed on stilts and using its tentacles to capture passing food. Unfortunately, *Hallucigenia* was reconstructed upside-down: the stilts are actually spines on the back of the body and the tentacles are limbs! Onychophorans have been traditionally viewed as a missing link between annelids and arthropods. Their bodies are certainly worm-like in shape but they grow by moulting as in arthropods. Molecular studies, however, suggest that onychophorans are not closely related to annelids and that their annelid-like features result from convergent evolution. Instead, together with another small phylum called the Tardigrada, they appear to be the closest living relatives of arthropods. Onychophorans probably made the transition to land sometime in the Palaeozoic, with all marine species subsequently becoming extinct.

## PRIAPULIDA

Another worm phylum, the Priapulida, seem also to have been more diverse during the early Palaeozoic than they are today. Only 17 species of these marine worms exist today but there are more than 30 species known from the early Cambrian to late Silurian before they almost disappear from the fossil record. Priapulids are predatory worms with cylindrical bodies up to 20 cm (8 inches) long. Their most striking feature is a structure called the introvert that can be turned inside out. At the apex of the introvert is the mouth surrounded by batteries of teeth. Priapulid worms hide in the mud on the sea bed and emerge to snatch their prey. The outer surface of the body is covered by a cuticle made of chitin.

**Above** A living onychophoran or velvet worm.

**Above** *Protoscolex*, a Silurian fossil from Shropshire, England, interpreted as a priapulid worm. The curved body is about 7 cm long.

As in arthropods and onychophorans, this cuticle is moulted during growth. It has recently become apparent that some Palaeozoic worms known as palaeoscolecids are almost certainly priapulids. Fragments of the tough cuticle of palaeoscolecids have a distinctive surface patterning formed by transverse bands of small nodes called papillae.

## SIPUNCULA

The Sipuncula ('peanut worms') resemble priapulids in many respects, particularly in having an introvert. However, whereas the gut of priapulids is straight, with the anus at the opposite end of the worm to the mouth, that of sipunculans is U-shaped. In contrast to the predatory priapulids, sipunculans are detritus or deposit feeders. Some of the 250 or so modern species occupy vacant gastropod shells in the manner of hermit crabs. Others live in sediment or bore into shell, coral and limestone substrates. Although some boring trace fossils are undoubtedly the work of ancient sipunculans, the body fossil record of this phylum is extremely meagre; only three or four occurrences have been described. The most celebrated of these occurrences is in the Silurian rocks of Illinois where aggregations of *Lecthaylus gregarius* have been found on bedding planes.

## ENIGMATIC TUBULAR FOSSILS

### TENTACULITIDS

These small Ordovician–Carboniferous fossils consist of elongate calcite tubes closed at the narrow end and with an opening or aperture at the broad end. Tentaculitid tubes are usually straight but are curved in some species. Fully-grown examples range from 4–30 mm (0.1–1 inch) in length, depending on species, and are generally about 1 mm (0.04 inches) wide at the open end. The outer surface is marked by transverse annulations, comprising prominent rings with gentler ridges called annulets in between, as well as a more subdued ornament of longitudinal ridges called lirae. Internally, up to 20 cross partitions (septa) can occur in the narrow, younger parts of the tube. A drop-shaped structure is present at the

**Above** Rare example of a fossil sipunculan worm. This 4 cm long specimen of *Lecthaylus* comes from the Silurian of Illinois.

**Above** Mass of small tentaculitid tubes, each less than a millimetre in diameter, from the Devonian of Ontario, imaged using a scanning electron microscope.

apex of the tube. This represents the initial skeleton which, by comparison with similar structures in modern invertebrates, was probably formed when a free-swimming larva metamorphosed into the adult.

Major questions remain to be solved about the mode of life of tentaculitids and also their relationships to other invertebrates. Species with thick-walled tubes seem – from their restricted occurrence in rocks deposited in shallow water, well-oxygenated environments – to have been benthic animals. Some examples of these thick-walled species have been found preserved in apparent life orientation, the tube vertically oriented with the apertural end uppermost. It seems likely that the living animal had the narrow end of its tube buried in stiff mud, the aperture protruding a little way above the sea bed. The majority of benthic tentaculitids are found lying horizontally and out of life position, having been disinterred from the sediment by current action. Dense accumulations of tentaculitid tubes in this orientation can be found in some deposits. In contrast to the thick-walled species, thin-walled tentaculitids were more widely distributed, being also found in deep water environments, and were probably planktonic animals. Both thin- and thick-walled species are most likely to have been suspension feeders consuming plankton. As for their phylogenetic affinities, the two main theories are that they are either a type of mollusc or are related to the brachiopods, bryozoans and phoronid worms, which together constitute a larger grouping called the lophophorates. The microstructure of tentaculitid tubes certainly resembles that found in many brachiopods and bryozoans, while it is not difficult to envisage the elongated body of some kind of worm fitting within the tube.

## CORNULITIDS

Cornulitids resemble tentaculitids on the one hand and serpulid worms on the other. Like tentaculitids, the calcite tubes of these Ordovician–Carboniferous fossils typically have a strong annular ornament. However, at least during their initial growth, cornulitids are cemented to a hard substrate and display an irregularity in growth that is much more reminiscent of serpulids than tentaculitids. Rather than signalling a phylogenetic affinity, the similarities between cornulitids and serpulids may well be due to convergent evolution reflecting adaptation to the same lifestyle. Both groups usually encrust shells or other hard substrates and, like some serpulids, cornulitids are not infrequently found growing towards the shell gape where they may have benefited from access to planktonic food entrained by the feeding currents of the host animal. Aggregations of cornulitids are also sometimes found, comprising closely packed tubes growing upwards from basal parts encrusting shells.

**Above** A cluster of cornulitids, 3.5 cm high, collected from the well-known Silurian locality called the Wren's Nest, Worcestershire, England.

**Above** Measuring 2 cm long and collected in Utah, this conical fossil is a specimen of the Cambrian hyolith *Haplophrentis*.

## HYOLITHS

The shells of hyoliths are found in Cambrian to Permian rocks. Several hundred species have been described, the majority coming from the Cambrian. The shell is thought to have been made of aragonite and was secreted from within by the soft parts of the animal. The main part of the shell is called the conch. This is conical, generally with a flattened (presumed) underside and a convex upperside, and is usually 15–30 mm (0.5–1 inch) long, although an exceptional specimen is known that is over 40 cm (15 inches) long. The narrow end of the conch is closed, the broad end open as an aperture. However, the aperture can be sealed by an operculum. A pair of peculiar whisker-like structures called helens extend out from between the conch and operculum in some species of hyoliths. The helens may have acted to stabilise the living animal on the sea bed and/or have been used to enable movement. Some fossil hyoliths have been found with sediment fillings of the gut. These show that the gut was basically U-shaped, though very sinuous in detail, and also suggest that hyoliths were deposit feeders that consumed large volumes of sediment to extract small organisms and organic debris. Possibly the living animal ploughed a shallow furrow in the sea bed, scooping up and feeding on the mud as it went. An alternative hypothesis views hyoliths as suspension feeding animals resting more or less stationary on the sediment surface and obtaining nutrition from plankton in the sea water.

Some palaeontologists consider hyoliths to be an extinct class of molluscs, an opinion based particularly on the structure of the shell that shows similarities in microstructure to molluscs. However, others believe that they were worm-like animals, perhaps related to the sipunculans.

# 5

# Jointed-limbed animals

NO OTHER PHYLUM of animals comes close to matching arthropods in terms of species richness. Arthropods are an enormously diverse and versatile group of animals with representatives inhabiting most types of environments, often in massive numbers. One subgroup of arthropods, the insects, alone account for perhaps 10 million living species. To these must be added many additional abundant and diverse kinds of arthropods, including spiders and scorpions, shrimps, crabs and lobsters, millipedes and centipedes, plus the extinct trilobites and eurypterids.

The secret of arthropod success lies in a body plan that is both efficiently 'engineered' and highly adaptable. Arthropods have bodies protected by exoskeletons made of two tough materials – chitin (a polysaccharide) and scleroprotein – sometimes further reinforced by a mineralised layer. The armoured exoskeleton not only serves to protect the animal but also acts as a surface on which the muscles are anchored and can work. Unlike many other invertebrates, arthropods lack the hydrostatic coelom that, through muscular action, permits the external shape of the animal to change for locomotion,

**Left** Complete specimen of the Welsh Ordovician trilobite *Ogygiocaris*, measuring 3.8 cm in length and showing the three-lobed dorsal exoskeleton.

burrowing and other purposes. Instead, the stiff arthropod body is jointed and contraction of various sets of muscles enables the individual units of exoskeleton to move relative to one another at the joints between them. This articulation is evident, for example, in the legs of a crab which typically have four or more joints. Tubular canals may penetrate the cuticle, linking to external sensory organs.

Arthropod bodies are conspicuously segmented. They are also regionalised into parts comprising series of identical or very similar segments. Particular regions, or tagmata, of the body are given different names in different kinds of arthropods. For instance, trilobites have a cephalon, thorax and pygidium, spiders a prosoma and metasoma, and insects a head, thorax and abdomen. The number of segments present in each region, as well as the total number of segments in the entire animal, vary according to the group in question, and sometimes within a single major group. Fusion of the segments is often evident. Primitively, each segment has a pair of appendages, one on the left and one on the right side. However, appendages are commonly missing from particular segments. As noted above, arthropod appendages are jointed. They are also modified in various ways to enable them to perform a range of functions, including walking, swimming, burrowing and processing food. Among the latter are hard jaw structures – mandibles – used for biting, cutting and grinding food.

The body cavity of arthropods is in large part occupied by a blood filled haemocoel that bathes the tissues of the animal, supplying them with oxygen and carrying away waste material. However, some arthropods, such as lobsters, also have organised circulatory systems with blood vessels. Respiration is accomplished in various ways. Myriapods and insects, which live mostly on the land surface, have a tracheal system comprising tubes (tracheae) for air breathing. Tracheal respiration becomes increasingly ineffective as body size increases and inner parts of the body fall beyond the reach of passive gaseous diffusion, an important factor limiting the maximum size attainable by terrestrial arthropods and explaining why this diverse group has never evolved species of elephantine or dinosaurian dimensions. Another group with terrestrial representatives, the arachnids employ so-called 'book lungs' for breathing air. These are modified versions of the gills used by many aquatic arthropods. Small aquatic arthropods, with large surface areas relative to their bodily volumes, simply allow gaseous exchange over the entire outer body surface.

Arthropod classification has long been a contentious issue for zoologists. It was once widely believed that 'arthropods' were polyphyletic, that is, they evolved several times from different ancestors, but this idea has largely fallen out of favour with the availability of better morphological analyses as well as molecular data. The inter-relationships between the major subgroups of arthropods are also becoming better understood, although they are yet to be finally settled. Five major subgroups of arthropods can be recognised: (1) trilobites (an extinct, Palaeozoic group); (2) chelicerates (e.g. horseshoe crabs, spiders, scorpions, eurypterids); (3) crustaceans (e.g. shrimps, crabs, barnacles, wood lice); (4) myriapods (centipedes and millipedes); and (5) insects (e.g. dragonflies, ants, beetles, wasps, flies). A close relationship between myriapods and insects – to

form a group called the uniramians – was once generally accepted but is now being challenged by new molecular evidence. It is probable that uniramians are united with the crustaceans. Trilobites are likely to be primitive, with chelicerates fitting somewhere between them and the crustaceans.

## TRILOBITES

Trilobites are endearing and attractive fossils. Not surprisingly, they are highly treasured by fossil collectors and there have even been attempts to promote the purchase of rare trilobites for investment purposes, as alternatives to postage stamps or gold. The name trilobite refers to the threefold (tri-lobed) lengthwise division of the body into an axial lobe flanked on either side by pleural lobes. Trilobite bodies are also divided at right angles to these lobes into three parts: cephalon (head), thorax (trunk) and pygidium (tail). The cephalon is made of segments fused into a single plate, as is the pygidium. Segments of the thorax, however, are not fused but hinged to their neighbours allowing for flexure. Like modern wood lice, which are crustaceans, many trilobites had the ability to roll themselves up into a ball; as a result of hinging, the thorax was able to bend through more than 180 degrees, bringing the head and tail into close contact. Examples of enrolled specimens are not uncommon in some species of trilobites. They undoubtedly represent a defensive posture that protected the vulnerable underside of the animal when disturbed or attacked by predators.

Over 15,000 species of trilobites have been named. They range in time from the early Cambrian to the late Permian. Thus, trilobites had the distinction of being players in two of the most momentous events in the evolutionary history of life, the 'Cambrian Explosion', when trilobites were among the many different sorts of animals making a debut in the fossil record, and the huge mass extinction at the end of the Permian that witnessed the final demise of trilobites. Between times, the diversity of different subgroups of trilobites waxed

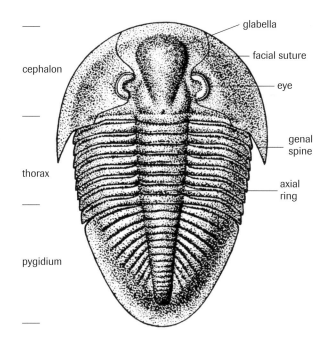

glabella

facial suture

eye

cephalon

genal
spine

thorax

axial
ring

pygidium

**Above** Labelled trilobite, the Ordovician genus *Basilicus*, illustrating
the important components of the dorsal exoskeleton.

and waned as different species appeared and became extinct. The late Cambrian and Ordovician were halcyon times for trilobites, numbers declining into the Silurian and more especially the late Palaeozoic, although some of the most striking trilobite species actually occur in the Devonian. The trilobite faunas of the Carboniferous and Permian do not match those of earlier times, not only in terms of the numbers of species present but also the range of variation, despite the fact that new species continued to evolve until the very end.

The majority of trilobites are just a few centimetres in length. However, the smallest species grew no bigger than 1 mm (0.04 inches) long, while the largest trilobite discovered, from the Ordovician of Manitoba in Canada, measured fully 72 cm (28 inches) in length. The reason that trilobites are common in the fossil record, at least relative to other arthropods, is that their dorsal exoskeletons were reinforced by the resistant mineral calcite. This is usually organised into two layers: an outer layer comprising prismatic crystallites of calcite

and a thicker, inner layer of microcrystalline structure. The thickness of the dorsal exoskeleton varies between species, being as little as 5 microns in some species and as much as 1 mm (0.04 inches) in others. Narrow tubular canals pass through the calcitic exoskeleton, some probably connecting with tiny sensory hairs on the outside of the animal. Only the dorsal (upper) part and the outer edges of the ventral (underside) part, called the doublure, of the trilobite exoskeleton are reinforced by calcite. Unfortunately, the limbs on the ventral side of the animal lack calcitic reinforcements and are seldom fossilised. In fact, the limb structure is well-known in very few genera of trilobites. Where preserved, limbs originated on all or most of the segments of the body. A pair of limbs at the front of the head were long and functioned as antennae. The other limbs are biramous, that is they had a branched structure, with an endopod and, closer to the main body of the animal, an exopod. The endopod was employed in locomotion and feeding, while the exopod supported a feathery structure believed to be a gill used in respiration and/or a sieve for straining small food particles.

Internal anatomy has been revealed by X-ray study of shales containing trilobites in which soft parts have been preserved through pyritisation, contrasting in density with both the exoskeleton and the surrounding rock matrix. The digestive system comprises a J-shaped oesophagus that leads forward from the mouth beneath the cephalon before bending backwards and passing into a short but broad stomach. After the stomach there is a long, narrow intestine, in some cases with blind lateral branches, that terminates in the anus situated beneath the final segment of the pygidium.

It is the dorsal exoskeleton only that is fossilised in most trilobites. The intricate structure of the dorsal exoskeleton, with substantial variations between different species, permits trilobite species to be more readily classified than is the case in many other invertebrate fossils. The cephalon or head of trilobites is particularly important in this respect. This is an approximately semicircular plate along the axis of which is a swollen structure called the glabella. Although it is tempting from its position to suppose that the glabella is some kind of braincase, this is incorrect. Instead, it was probably a covering for the stomach that lay directly beneath. Lineations on either side of the cephalon outside the glabella are called facial sutures. These mark the boundary between the inner part of the cephalon that was retained when the animal moulted, and the outer part (free cheeks) that was shed during moulting.

Most trilobites have crescent-shaped eyes located along the facial sutures on either side of the glabella. The exoskeleton of the eyes was moulted with the free cheeks. Trilobite eyes have been claimed to be the most ancient evidence for vision in animals. As in many other arthropods, these are compound, with numerous individual lenses of small size. A lot is known about the visual system in trilobites because the lenses, being modifications of the normal exoskeleton, are made of calcite which fossilises very well. Two distinct types of eyes occur in trilobites: most species have so-called holochroal eyes, but species belonging to the suborder Phacopina have schizochroal eyes. Holochroal eyes generally have a large number of lenses, occasionally up to 15,000, which are tightly packed together. As is commonly the case with tightly packed structures, each lens has a hexagonal outline shape. A single corneal membrane covered all of the lenses in the holochroal eye. In some species the lenses are shallow and biconvex, like a magnifying glass, but in others they are long and prismatic. Each lens is a single, precisely oriented crystal of calcite, able to focus sharply the light passing through it onto sensory receptors that would have been present at the rear of the eyes. Holochroal eyes in trilobites are occasionally set on stalks.

The second type of eyes – schizochroal – have a smaller number of lenses that are large, thick, roughly circular in outline, separated from neighbouring lenses and furnished with their own corneal membranes. The individual lenses of schizochroal eyes are complex, being

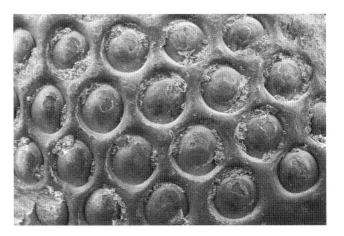

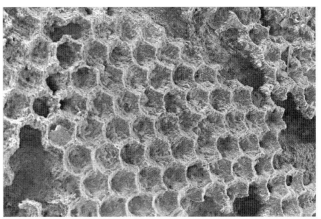

**Above** Detail of schizochroal (upper fig.) and holochroal (lower fig.) trilobite eyes. These scanning electron micrographs show the eye lenses magnified by x 30 and x 50 respectively.

constructed of two units of calcite that together produced a well focused image free of astigmatism. Elegant research on the functioning of trilobite eyes has shown not only how the optics of vision worked, but has also allowed estimates of the visual fields possessed by different species. The latter is important when trying to understand the mode of life of trilobites, as discussed below.

The eyes were lost during the evolution of trilobites belonging to several different groups. In one evolutionary lineage of Devonian trilobites it is possible to trace the progressive shrinking of the eye, culminating in its total disappearance. Eye loss is often taken to indicate that the animals became adapted to life in deep water where light levels were negligible and other senses were of greater importance. However, even shallow water trilobites sometimes lost their eyes.

Other notable features of the cephalon are the cephalic fringe and spines. The cephalic fringe is situated around the curved front edge of the cephalon. It is present only in some trilobites having no or reduced eyes and with facial sutures at the very margins of the cephalon. Rows of pits in the cephalic fringe are thought by some palaeontologists to have housed sensory hairs able to gauge the flow of water over the surface of the animal. However, the function of these pits has yet to be settled. The two corners of the cephalon, known as genal angles, are often the sites of genal spines that sweep back towards the pygidium. In some species genal spines are very long, at least twice the length of the main body of the animal. Occasionally, a spine projects forward from the front end of the cephalon. Several different functions may have been served by spines, perhaps the most prevalent being to deter predators.

The thorax of trilobites is less variable than the cephalon. However, the number of thoracic segments does vary between species, from two to over sixty. Segment size tends to decrease slightly towards the pygidium, and in some species the outer edges of the segments (pleura) have spine-like prolongations. Along the central axis of the trilobite the thoracic segments are axial rings that may bear median spines. Between one and thirty segments are fused together to form the pygidium. This is probably the most commonly collected part of a trilobite because it is relatively resistant and immediately recognisable. Occasionally, the pygidium ends in a long spine.

A pattern of parallel terrace ridges is sometimes visible on the dorsal exoskeleton. Each terrace ridge has a steep face and a shallow slope leading back to the steep face of the next concentric ridge, like a series of escarpments. Similar terrace ridges are known in living arthropods, such as crustaceans, where they may serve to assist in burrowing, the steep face of the ridge 'gripping' the sediment against the force applied by the animal. This gripping ability also makes it harder for the animal to be pulled out of its burrow by a predator.

Like other arthropods, trilobites moulted. Indeed, the majority of fossil trilobite specimens are cast-off parts of moulted exoskeletons; fossils of complete animals that died and were buried in an intact condition are much less common. Discoveries of populations of trilobites consisting of individuals ranging from juveniles to adults has allowed palaeontologists to study how the dorsal exoskeleton changed during growth. Successive moult stages (instars) show, for instance, that more and more segments were added to the thorax until the adult condition was achieved, often at quite a small size. Moulting continued through adulthood but segment number and other features of the skeleton, apart from size, remained much the same from one instar to the next.

Understanding the mode of life of trilobites presents a major challenge to palaeontologists as they have no close living relatives with which to draw analogies. Making generalisations about trilobite ecology is also unwise: while trilobites have a characteristic body plan, variations between species in the structure of the exoskeleton are large enough to imply that different trilobites enjoyed widely different lifestyles. Furthermore, their occurrence fossilised in various kinds of sediment – from shallow water sandstones, through limestones, including reefs, to deep-sea shales – points to the fact that they could colonise many different environments. One thing is certain, all trilobite species lived in the sea; there is no evidence of freshwater trilobites, let alone terrestrial species. The best evidence for trilobite lifestyles comes from detailed analyses of the skeleton coupled with interpretation of their geological occurrences. Eight main morphologies of trilobites have been recognised, of which five can be equated with particular ecologies: (1) 'pelagic' trilobites with enlarged, sometimes bulbous, eyes affording very wide fields of view swam in shallow to deep water environments; (2)

'atheloptic' trilobites also lived in deep waters but these were blind or had reduced vision; (3) 'phacomorph' trilobites with a large glabella and furrowed pygidium colonised shallow water environments where limey sediment was being deposited; (4) 'illaenimorph' trilobites having smooth-surfaced exoskeletons often, though not always, inhabited reefs; (5) 'olenimorph' trilobites with thin skeletons and numerous narrow thoracic segments lived in environments with low levels of oxygen.

While some trilobites were undoubtedly swimmers, others crawled over the sea bed, and yet others burrowed just beneath the surface of the sediment. The tracks of crawling and shallow burrowing trilobites are sometimes evident on bedding plane surfaces, in rare instances with the maker preserved in association with the trace fossil. A trace fossil genus called *Cruziana* was usually,

but not always, made by trilobites. These gently meandering trails have round-topped ridges separated by a central furrow and bear chevron markings, made by the legs, with the V-shape opening in the direction that the animal moved.

Most trilobites probably lived with their bodies oriented approximately parallel to the sea bed, whether on the surface or just beneath in a shallow burrow. However, the Ordovician species *Ptyocephalus vigilans* was an exception. A discovery consisting of several specimens distributed over a small area of bedding plane in life position has shown that the cephalon was oriented horizontally on the sea bed, whereas the thorax and pygidium were buried vertically in the sediment.

It seems very likely that trilobites employed a variety of different feeding modes depending on the species. Some were probably scavengers or carnivores capable of

## TRILOBITE ORDERS

- Agnostida: tiny trilobites with only two or three thoracic segments; early Cambrian–late Ordovician; e.g. *Eodiscus*.

- Redlichiida: primitive trilobites with numerous thoracic segments that are usually spiny, large eyes and a small pygidium; early–mid Cambrian; e.g. *Olenellus*.

- Corynexochida: typically spiny trilobites with ten or fewer thoracic segments; early Cambrian–late Devonian; e.g. *Bumastus*.

- Ptychopariida: a highly varied, generalised group of trilobites; early Cambrian–late Devonian; e.g. *Elrathia*.

- Harpetida: trilobites with large cephalons, very small eyes, numerous thoracic segments and a short pygidium; late Cambrian–late Devonian; e.g. *Eoharpes*.

- Proetida: including some of the youngest trilobites, proetids have a large, domed glabella and thorax of eight to ten segments; late Cambrian–late Permian; e,g. *Phillipsia*.

- Phacopida: another varied group of trilobites, difficult to demarcate, which includes the suborder Phacopina characterised by schizochroal eyes; early Ordovician–late Devonian; e.g. *Calymene, Dalmanites, Phacops, Trimerus*.

- Lichida: trilobites in which there is a broad glabella, large pygidium and the exoskeleton is ornamented by tubercles; mid Cambrian–late Devonian; e.g. *Odontopleura*.

- Asaphida: large, smooth trilobites with six to nine thoracic segments; mid Cambrian–late Silurian; e.g. *Asaphus*.

feeding on worms and other animals, possibly using spiny structures called gnathobases on the legs to capture and tear apart their prey and pass the pieces forwards towards the mouth. An enlarged glabella in species inferred to have been carnivores may reflect the existence of a large stomach beneath and the capacity to swallow relatively big prey animals. Presumed swimming trilobites with large eyes like *Carolinites* may have been visual hunters of other planktonic animals. Trilobites with large, domed cephalons, such as *Trinucleus*, are hypothesised to have been suspension feeders, perhaps using gill filaments on the limbs to create a feeding current and to capture the small particles of food as they were swept past. Deposit feeding, in which nutrients are extracted from sediments processed through the gut, is another feeding mode found commonly among modern arthropods and in all probability also adopted by some trilobites.

There can be little doubt that trilobites themselves were eaten by other animals. Not infrequently trilobites are found with damaged and repaired skeletons suggestive of attacks by predators. Only very recently, however, has more direct evidence of predation on trilobites come to light with the discovery of fragments of trilobites in the gut contents of an unknown predatory animal from the Middle Cambrian of China.

Depending on the classification used, eight or nine orders of trilobites are recognised but the inter-relationships between different trilobites are still matters of scientific debate.

## *BUMASTUS*: SILURIAN OF EUROPE, NORTH AMERICA AND AUSTRALIA

*Bumastus* is a large trilobite with a smooth exoskeleton about twice as long as wide. The cephalon is approximately two times wider than long and has eyes of moderate size set well towards the back, a subdued glabella with a median tubercle, and rounded genal angles. There are ten narrow thoracic segments. The pygidium is rather featureless.

**Above** A 10.7 cm long specimen from the British Silurian of the trilobite *Bumastus*.

It has been suggested that *Bumastus* burrowed backwards into the mud, leaving the eyes peeping above the sediment surface.

## *CALYMENE*: SILURIAN–DEVONIAN, WORLDWIDE

Up to 10 cm (4 inches) long, this convex trilobite has a swollen glabella with deep furrows. Genal spines are absent and the facial sutures run diagonally across the cephalon. The eyes are small and schizochroal. The gently tapering thorax has 13 segments, turned down steeply at their outer edges. Shorter than the cephalon,

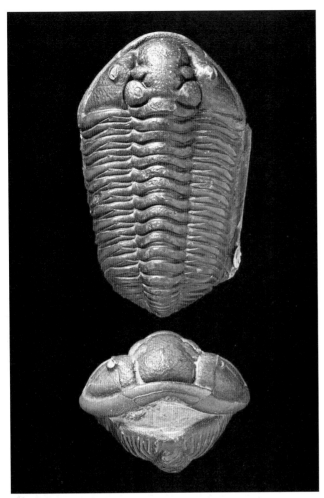

**Above** Normal and enrolled examples of *Calymene*, about 2.5 cm wide, from the Silurian Wenlock Limestone of England.

distance along the sides of the thorax, which has 11 segments. The pygidium has a triangular shape, terminating in a spine that can be moderately long.

Judging from the size, shape and prominence of the eyes, *Dalmanites* had good vision, particularly out to the sides of the animal.

### *ELRATHIA*: CAMBRIAN OF NORTH AMERICA

This small trilobite is oval in plan view and low in profile. The glabella and eyes are small, and short genal spines are present. Thirteen segments make up the thorax, which has a narrow axial region compared to the pleura on either side. The pygidium has a well rounded termination, is of moderate size, and measures about half as long as wide.

Flat, disc-shaped specimens of this trilobite occur in tremendous profusion through a small interval of the mid Cambrian Wheeler Shale in Utah. It is believed that these trilobites were able to inhabit environments low in oxygen that excluded most other animals. Specimens are often to be found for sale in rock and fossil shops, sometimes mounted in jewellery or backed by a magnet for sticking onto a refrigerator.

the pygidium comprises about seven fused segments. Tightly enrolled specimens of this trilobite are often found.

*Calymene* lived in shallow water environments and was probably a predator.

### *DALMANITES*: SILURIAN–DEVONIAN, WORLDWIDE

*Dalmanites* (see colour fig. 24) is a relatively flat trilobite. The low glabella expands in width towards the front end of the cephalon and has three pairs of lateral furrows. The scizochroal eyes are large, strongly curved and upstanding. Genal spines are present, running for some

**Above** The small Cambrian trilobite *Elrathia* represented by a 2 cm long specimen from the House Range of Utah.

**Above** The spiny trilobite *Leonaspis* exemplified by a 12 mm long individual from the Silurian of Bohemia.

## *LEONASPIS*: SILURIAN–DEVONIAN OF EUROPE AND NORTH AFRICA

This spiny trilobite has a short, wide cephalon with long, slightly raised genal spines that extend backwards and outwards. A further pair of long, horn-like spines extend upwards and backwards from the axis of the cephalon close to its border with the thorax. The glabella is lobate, and the eyes small. All nine segments of the thorax have long marginal (pleural) spines. The main part of the pygidium is short but there are several spines around the margin, of which one pair is longer than the rest.

Many Silurian and Devonian trilobites are highly spinose. The most reasonable functional interpretation of this feature is to discourage predation. Indeed, the Devonian is regarded by some palaeontologists as a time when levels of predation in the sea increased dramatically, with important new groups of predatory animals first appearing and their prey species responding by evolving defences such as spines.

## *OLENELLUS*: CAMBRIAN OF NORTH AMERICA AND NORTHERN EUROPE

Up to about 15 cm (6 inches) long, this primitive trilobite lacks facial sutures. The cephalon bears crescent-shaped eyes, joined to the frontal lobe of the glabella, and genal spines. There are 14 segments in the thorax, the third of which is noticeably larger and more spinose than the others. The small pygidium is elongate.

The olenellids are the earliest trilobites recorded from the fossil record in North America and Scandinavia. While some authors have argued that they are not true trilobites, partly on account of the peculiar enlargement of the third thoracic segment, most evidence favours their inclusion within the trilobites.

**Above** An incomplete specimen of *Olenellus*, 1.3 cm wide, showing the enlarged third thoracic segment. The trilobite was collected in Scotland from rocks of Early Cambrian age.

## *PHACOPS*: DEVONIAN, WORLDWIDE

A medium-sized trilobite with a convex dorsal exoskeleton. The cephalon lacks genal spines and has a glabella that expands in width forwards, lacks incised furrows and is generally covered by tubercles. The gently curved schizochroal eyes have a small number of large lenses. There are 11 thoracic segments, deeply inturned at the edges. The pygidium is semicircular with a tapering axis comprising six to eight axial rings.

This common trilobite can be found in limestones deposited in shallow water, for example in Morocco and New York State. The schizochroal eyes are a distinctive feature and it is possible that the animal had binocular vision enabling it to hunt prey in low light levels. An entirely different interpretation, however, views *Phacops* as an algal grazer.

**Above** This Moroccan Devonian *Phacops* is enrolled, measures 4.5 cm in width, and has a glabella covered in tubercles.

## *REMOPLEURIDES*: ORDOVICIAN, WORLDWIDE

A small trilobite with a wide axis but narrow pleural regions. Most of the cephalon is occupied by the large glabella flanked by the large, narrow eyes with minute lenses. There are no genal spines. The barrel-shaped, tapering thorax has 11 segments. Only one or two segments comprise the pygidium.

This trilobite is thought to have been a free-swimming member of the plankton, using the large eyes to hunt its prey visually. Such a mode of life allowed the animal to become geographically very widespread in Ordovician times.

**Above** Probably planktonic, *Remopeurides* is a small trilobite – this Scottish Ordovician example being less than 2 cm long – with an inflated glabella.

## *TRIMERUS*: SILURIAN–DEVONIAN, WORLDWIDE

This often large trilobite has a vaulted, elongate dorsal exoskeleton with a smooth surface topography. The cephalon is subtriangular, about the same length as the pygidium, both being considerably shorter than the thorax. Genal spines are lacking, the glabella has no furrows, and the eyes are small, raised and not in contact with the glabella. The thorax has 13 segments and a broad but indistinctly demarcated axis. Subtriangular in outline, the tapering pygidium is arched and has six or more transverse furrows.

*Trimerus* with its smooth, streamlined body is thought to have been a shallow burrower, the raised eyes protruding above the sediment surface.

**Right** Fully preserved, 10 cm long specimen of the Silurian trilobite *Trimerus*.

## TRINODUS: ORDOVICIAN OF NORTH AMERICA AND EURASIA

A tiny elliptical trilobite often about 5 mm (0.7 inches) in length. The cephalon and pygidium are both relatively large and of superficially similar appearance, whereas the

**Above** The pitted cephalic fringe is clearly seen in this 1.3 cm wide specimen of *Trinucleus* from the Ordovician of Wales.

**Above** Measuring less than a cm, this Scottish Ordovician trilobite *Trinodus* has a much reduced thorax between the larger head and cephalon.

short intervening thorax comprises only two segments. There are no eyes or genal spines, and the glabella is spindle-shaped with a faint glabellar furrow. The pygidium has a tapering axis with few rings.

This blind trilobite may have lived in the plankton.

## TRINUCLEUS: ORDOVICIAN OF EUROPE AND ASIA

A small trilobite with genal spines exceeding the length of the rest of the elliptical or subcircular body. The cephalon is large, accounting for almost half of the length of the body, and convex, with a wide fringe having four to seven arcs of pits. Eyes are lacking. The glabella is inflated and pyriform in shape. Six segments comprise the thorax. The short, triangular pygidium has a smooth margin.

This blind trilobite is thought to have fed by filtering fine particles stirred up into suspension. The wide cephalon may have acted like a snowshoe in preventing the animal sinking into the mud.

# CHELICERATES

Chelicerates are a major group of arthropods that contains spiders, scorpions, mites and horseshoe crabs (which are not true crabs), as well as a fascinating extinct group called eurypterids. An estimated 65,000 species of chelicerates are known, most of these being spiders and mites of various sorts. A principal feature that distinguishes chelicerates from other arthropods is the division of the body into two regions (tagmata), a prosoma of six segments, equivalent to a fused head and thorax of other arthropods, and a opisthosoma generally of 12 segments, except for primitive species that apparently had only 11 opisthosomal segments. In addition, a pair of pincers arise from the first segment of the prostoma. These are called chelicerae and give the group its name. Despite several shared characteristics, the main chelicerate subgroups – Xiphosura, Eurypterida and Arachnida – are sufficiently different in morphology and ecology that they are best treated separately.

Xiphosurans are the so-called horseshoe crabs, sometimes also known as limulids. The good times are in the past for xiphosurans; although this ancient group first appeared in the fossil record in the early Palaeozoic, there are only four species still living today. Best known of these is *Limulus polyphemus* (see colour fig. 25), the common horseshoe crab of the Atlantic seaboard of North America that is often regarded as a 'living fossil'. This animal inhabits shallow marine waters, spending much of the day buried in the sand but emerging at night to feed on worms and molluscs. It is able to crawl ashore and also to swim upside down. *Limulus* has a vaulted, shield-like carapace beneath which there are six pairs of jointed appendages, each terminating in pincers, the five rear pairs being the walking legs. All of the appendages are capable of gathering food that they grind using structures called gnathobases, passing the pieces forward towards the mouth. Compound eyes are present on the top of the carapace. Behind the carapace is the opithosoma, comprising the abdomen and a long, spine-like tail. The total length of the animal can be up to 60 cm (24 inches), with large individuals tipping the scales at 5 kg (11 pounds).

The cuticle of xiphosurans can be tough but is thin and uncalcified, hence their fossil record is scanty. The oldest fossils displaying xiphosuran-like features date from the Cambrian, but more certain representatives first appear in deposits of Silurian age. A total of about 30 genera of xiphosurans are known in the fossil record, those from the late Palaeozoic onwards being very similar in morphology to the modern *Limulus*. Unlike *Limulus*, however, the majority of these fossil xiphosurans apparently inhabited freshwater or brackish environments. They also tend to be smaller in size. Some of the best fossil xiphosurans come from the late Jurassic Solnhofen Limestone of Bavaria where the genus *Mesolimulus* at first sight appears almost indistinguishable from the modern *Limulus*. There are even remarkable examples of individuals of *Mesolimulus* preserved at the ends of the tracks they made in the mud – Jurassic 'walks of death'.

The largest arthropod ever to have lived is the Silurian–Devonian eurypterid *Pterygotus*, which could reach 2.3 m (7.5 feet) in length. Although never particularly common as fossils and consisting only of about 50 genera, eurypterids or sea-scorpions are an Ordovician to Permian group of chelicerates that were remarkable for being the top predators in aquatic ecosystems through much of their duration. The eurypterid body comprises a relatively short and shield-like carapace (prosoma) on which the compound eyes are situated, followed by a tapering opisthosoma of 12 movable segments and a tail or telson that can be pointed or take the form of an oval plate. Borne on segments of both the prosoma and opisthosoma, the appendages vary in their shape along the length of the animal, reflecting variations in their function. The second to fifth pairs of appendages seem to have been adapted for walking, while the sixth pair of appendages are paddle-like and were very probably employed in swimming. It has been suggested that these swimming appendages made a figure of eight in the water, a pattern resembling rowing, to propel the animal forwards. In some eurypterid species it has been possible to distinguish between the sexes based on differences in a structure on the underside of the animal called the genital appendage. This can be either broad or narrow but it is not known which of the two conditions is male and which is female. Also on the underside of the body are covered chambers holding the gills. Covering of the gills may have allowed some eurypterids to emerge from the water and onto land for short periods without the gills drying out entirely.

Most eurypterid fossils come from sediments deposited in brackish or freshwater, although some occur in marine sediments. While it is possible that some species were able to inhabit waters of varying salinity, eurypterids show a general transition through time from being predominantly marine in the Ordovician to being inhabitants of non-marine environments during the later part of their evolutionary history. There was probably also considerable variation between species in how they moved. Judging from detailed studies of their appendages and associated tracks, some species were adept walkers on the sea bed, others crawled inelegantly and yet others got about mostly by swimming. It is highly likely that most or all eurypterids were carnivores. Possible coprolites (fossilised faeces) of eurypterids have been found to contain fragments of trilobites, graptolites, fish and even other eurypterids. Features of the animals themselves that point to a carnivorous lifestyle include large body size (invariably useful for overpowering prey), their apparently good stereoscopic vision, and the structure of the first pair of appendages that end in strong, pincer-like chelicerae.

Arachnophobia is popularly defined as a fear of spiders. However, arachnids are a group including not only spiders but also the equally fearsome scorpions as well as harvestmen, whip scorpions, mites and ticks. Although by far the most abundant and diverse

chelicerates today, arachnids are not well represented in the fossil record. Those examples that do occur tend to be fossilised in special circumstances, for example, by entrapment in tree resin that solidified to become amber, or in concretions where early mineralisation promoted survival of the entombed carcass. In arachnids the prosoma is covered by a carapace and bears most if not all of the appendages, the opisthosoma is often clearly segmented, except in spiders, and the eyes are simple, in contrast to the other chelicerates that have compound eyes. Most arachnids live on land, breathing using tracheae, so-called 'book lungs', or a combination of the two methods. Scorpions are the first arachnids to appear in the fossil record. Unlike modern scorpions, however, these Silurian examples appear to have possessed gills, pointing to an aquatic rather than a land-dwelling existence for the early members of the group. Not until the Carboniferous are scorpions found with book lungs that indicate adaptation to life on dry, or more likely moist land surfaces.

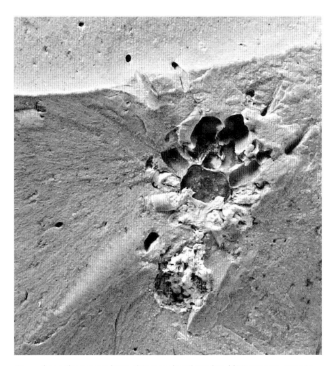

**Above** Less than 1 cm long, this rare fossil spider, *Vectaraneus*, comes from Oligocene deposits in the Isle of Wight, England.

Some remarkable examples of spider fossils have been discovered in recent years, including *Vectaraneus* from the Eocene of the Isle of Wight in southern England. In this water spider the muscles, book lungs and tracheal system have all been replaced by calcium carbonate and preserved in three-dimensional detail. A group of arachnids closely related to the spiders are trigonotarbids. These probably occupied some of the same ecological niches in Silurian to Permian times that spiders subsequently occupied. Trigonotarbids differ from spiders in several respects, notably in lacking the specialised appendages known as spinnerets that spiders use to produce silk for their webs, to wrap the bodies of their prey, and to swing through the air.

**Above** The paddles used for swimming are very conspicuous in the Silurian eurypterid *Eurypterus* from New York State. This individual measures 12 cm in length.

### *ESURITOR*: EOCENE–OLIGOCENE OF EUROPE

The pear-shaped carapace of this spider (see colour fig. 26) is broader than the opisthosoma (abdomen), which is unsegmented and bears six silk-producing spinnerets. There are four pairs of long, hairy legs with tiny terminal claws, the anterior legs having long spines. The anterior eyes are closer together than the posterior eyes. The chelicerae are fang-like.

*Esuritor* is one of numerous spider genera found in amber. This particular genus is from Baltic Amber that originates from a late Eocene–early Oligocene formation called the Blue Earth, coming to the surface on the bed of the Baltic Sea. Storms rip up pieces of amber from the Blue Earth and deposit it on the beaches around Kaliningrad in Russia and also along the Baltic coasts of Lithuania, Poland, Germany and Denmark.

### *EURYPTERUS*: ORDOVICIAN– CARBONIFEROUS OF NORTH AMERICA, EUROPE AND ASIA

This eurypterid can be up to 28 cm (11 inches) long. The prosoma has a rounded quadrate shape with slightly curved, compound eyes positioned slightly in front of the mid-point. A strongly expanded paddle is present on the swimming legs and the walking legs are mostly spinose with small cheicerae. In the presumed male the genital appendage is long and has diverging spines; in the presumed female it is short with covered lateral lobes. The telson is long and pointed.

*Eurypterus remipes* was adopted in 1984 as the State Fossil of New York State.

### *MESOLIMULUS*: JURASSIC–CRETACEOUS OF EUROPE

A vaulted semicircular carapace covers the prosoma of the xiphosuran *Mesolimulus* and has a median longitudinal ridge, two small eyes, and genal spines at the posterior angles. The opisthosoma comprises fused segments with pleural spines giving the outer edges a opisthosma comprises a wide mesostoma of seven segments and a longer and more slender metastoma of five segments, plus a telson (stinger) that may have had a poison gland. A genital operculum is present on the first mesostomal segment. The coxae (leg segments nearest the body) of the first two pairs of legs are enlarged for feeding.

This early scorpion probably lived in water for all or most of the time, unlike modern scorpions which are fully terrestrial.

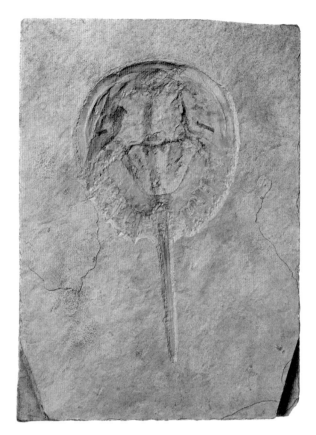

**Above** The Jurassic horseshoe crab *Mesolimulus*, 17 cm long, from the famous fossil locality of Solnhofen in Bavaria, Germany.

jagged appearance. The 'tail' (telson) is long and pointed.

This genus of horseshoe crab is particularly well-known from the famous Solnhofen Limestone of Bavaria, Germany. Exquisite examples can be seen in many museum collections. They owe their superb preservation to the particular environment of deposition of this fine-grained, 'lithographic' limestone, which is also the source of the early bird *Archaeopteryx*. The Solnhofen Limestone was apparently deposited in a tranquil lagoon, the sea floor comprising stagnant lime mud on which animals periodically became stranded.

## *PARAISOBUTHUS*: CARBONIFEROUS OF EUROPE AND NORTH AMERICA

This scorpion has a carapace with well-developed cheeks on either side of a deep groove and lateral eyes. The

**Above** *Paraisobothus*, an early scorpion from the Carboniferous of the Czech Republic, a little over 9 cm in length.

# CRUSTACEANS

Crustaceans have left a fossil legacy second only among arthropods to that of trilobites. This is because, like trilobites, many crustacean species employ calcium carbonate biominerals to reinforce their skeletons and are mostly aquatic and therefore more likely to survive as fossils than say spiders or insects. Crustaceans exhibit a remarkable diversity of morphologies and lifestyles. The most familiar crustaceans are lobsters, shrimps, crabs and barnacles. However, there are many other kinds and it is thought that more than 50,000 species of crustaceans are extant. While the majority of these species live in the sea, at depths ranging from the intertidal to the deep abyss, a significant number inhabit freshwater environments and a few, notably wood lice, live on land. Such is their diversity in marine environments that they have been dubbed 'the insects of the sea'. Crustaceans also encompass a vast range of body sizes, from tiny parasitic and sediment-dwelling species, to giant crabs with legs spanning 4 m (13 feet). They include herbivores, carnivores and scavengers. Among marine crustaceans some species are benthic, sea bed dwellers, either living on the surface or burrowing sometimes more than 1 metre (3 feet) down into the sediment, while others swim or float in open water. A vast biomass is represented by the shrimp-like krill that live in densities of up to 1000 individuals per cubic metre, particularly in the southern polar ocean.

A suite of shared features shines through the great variability of forms evident in crustaceans and allows them to be distinguished from other arthropods. All have a head composed of five segments followed by a long trunk usually clearly divided into a thorax and an abdomen. Crustacean larvae – referred to as naupilus larvae – are oval in shape and have three pairs of appendages: antennules that are unbranched (uniramous), antennae that are branched (biramous, as in trilobites) and mandibles. These appendages form part of the head in adults. It should be noted, however, that a lot of crustacean species lack naupilus larvae.

Adult crustaceans also develop additional appendages, including two extra pairs of head appendages called the maxillae, and a variable number of appendages on the thorax and abdomen that can be employed in feeding, walking or swimming. Different limbs along the body of an individual crustacean are commonly specialised to perform different functions. Compound eyes are usually present, sometimes situated on stalks, as in crabs. Often the body is flattened in one plane or the other, either laterally (as in shrimps) or dorso-ventrally (as in woodlice).

The fossil history of crustaceans extends back at least to the Cambrian. There has even been a disputed claim that a late Precambrian, 'Ediacaran' fossil called *Parvancorina* is a crustacean. Although more than 2000 crustacean genera are represented in the fossil record, this is not a particularly large number when the huge diversity of crustaceans that live today is taken into account. Indeed, with a few notable exceptions, crustacean fossils tend to be of localised occurrence and are not nearly as abundant as, for example, molluscs. The high organic content in their mineralised skeletons means that they are prone to disintegrate unless buried rapidly and protected from scavengers and micro-organisms causing decomposition. Resorption of calcium prior to moulting further weakens the skeleton and diminishes the likelihood of fossilisation.

From a palaeontological perspective, the most important crustaceans belong to the classes Malacostraca, Branchiopoda, Ostracoda and Cirripedia. The earliest crustaceans date back to the early Cambrian. Moderately common in the Palaeozoic are representatives of a primitive subgroup of crustaceans called the phyllocarids, probably close relatives of malacostracans, which have bivalved carapaces covering the front part of the body of 19 segments. Phyllocarids are widespread marine animals, many inhabiting muddy sea floors but some swimming in open waters. While some phyllocarids were filter-feeders, creating a water current to bring small food particles towards the mouth, large species of the Silurian-Devonian genus *Ceratiocaris*, which exceptionally reach a length of 75 cm (29 inches), probably used their mandibles to cut up bigger prey. Although phyllocarids still exist today, the vast majority of modern malacostracans belong to other subgroups, among which reside the shrimps, lobsters, crabs, hermit crabs, isopods and amphipods. The first four of these are members of the Decapoda, the name signifying that they have ten legs.

The oldest examples of decapods come from the Triassic. However, related shrimp-like forms date back to the Devonian and, given that they have non-mineralised skeletons made of chitin that do not fossilise well, it is quite possible that they have a more ancient history not represented in the fossil record. Shrimps are normally only fossilised when the effects of decay are countered by early growth of diagenetic minerals; they are often found in phosphatic or carbonate concretions formed under such circumstances. Analysis of appendage structure shows that late Palaeozoic 'shrimps' were predominantly carnivores, many species living in brackish water environments. Lobsters and crayfish are typically large decapods with strong tail fans. They can inhabit both marine and freshwater environments. Like crabs, lobsters have pincers that can grasp their prey, an ability that has doubtless contributed to their success.

Crabs are better represented in the fossil record by virtue of their strongly calcified carapaces and chelae (claws). In true crabs (Brachyura) the abdomen is folded beneath the carapace of the thorax, which tends to be broad and flat. Lifestyles vary among species of brachyuran crabs. Most live in the sea but some inhabit freshwater and a few are terrestrial. They can be carnivores, herbivores, omnivores, suspension feeders or deposit feeders. Their dominant mode of locomotion can be placed in one of five main categories – swimming, walking, running, climbing, burrowing – reflected in the fossilisable hard parts. In swimming crabs, the last pair of appendages are flattened to form paddles. Walking crabs have sturdy appendages and heavy shells whereas running crabs are more lightly constructed. Terrestrial crabs that are able to climb trees also have thin shells but their limbs end in short, sharply pointed dactyli allowing them to grip rough surfaces. Finally, burrowing crabs are of two kinds, side-burrowers and back-burrowers. Side-burrowers have flattened dactyli on the final legs that are used for digging, while back-burrowers are usually smooth and flat, allowing them to become immersed easily into the sediment.

Appearing initially in the Jurassic, crabs were rare until the Cretaceous when they underwent a major evolutionary radiation. Crab chelae and carapaces can be moderately common fossils in Cretaceous and Cenozoic limestones. Occasional spectacular examples of more

**Above** Beautifully preserved specimen, 6 cm wide, of the crab *Lobonotus* from the Oligocene of Haiti.

complete fossil crabs are found in concretions from fine-grained marine sedimentary rocks such as mudstones and shales. As always, rapid burial promotes their fossilisation; exoskeletons of dead crabs or moults left on the sea bed may contain sufficient organic material to be attractive, both to large scavenging animals and to bacterial decomposers. Some of the best preserved fossil crabs represent individuals buried alive in their burrows; these have a characteristic pose, with the claws held against the edges of the carapace and the legs folded beneath the body.

The hermit crabs (Anomura) typically inhabit the shells of dead gastropods. They also appeared in the Jurassic but have a poorer fossil record than the true crabs. Only a handful of examples are known of fossil hermit crabs preserved *in situ* in their gastropod shell homes. Sometimes, however, it is possible to infer the former presence of a hermit crab in a fossil gastropod shell. Hermit crabs do not look after their shell homes as well as the gastropods that originally made them, and the shells may acquire tell-tale worn patches indicative of dragging across the sea-floor by their crab occupant. A variety of encrusting animals may also colonise the surfaces of gastropod shells tenanted by hermit crabs. Among these are sponges, hydrozoans

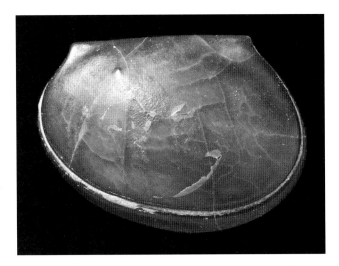

**Above** By the standards of ostracods, the Silurian genus *Leperditia* is enormous, this individual measuring 14 mm in width.

and bryozoans that not only develop very thick colonies but also extend the shell by growing outwards around the shell aperture, thereby providing the hermit crab with a home that is less likely to be crushed by predators and which grows in concert with the crab.

Two other malacostracan groups – amphipods and isopods – are of great importance at the present day, each containing about 10,000 living species, but have scant fossil records. Amphipods resemble shrimps, but lack a carapace, and are found in marine and freshwater environments. Isopods also lack a carapace, and live in the sea, freshwater or on land (e.g. woodlice). They are slightly better represented in the fossil record than amphipods, the Jurassic–Cretaceous genus *Archaeoniscus* being an example of a fossil isopod that can be locally abundant.

Branchiopods, not to be confused with the totally unrelated brachiopods, are aquatic crustaceans that include the so-called fairy shrimps and water fleas such as *Daphnia*. One subgroup of branchiopods with palaeontological significance is the Order Conchostraca. Conchostracans have a carapace of two valves that completely encloses the body. This carapace is ornamented with growth lines and so closely resembles a bivalve molluscs that conchostracans have received the vernacular name "clam shrimps". Unlike bivalve molluscs the valves of conchostracans are not calcified. There are about 200 described species of conchostracans living at the present day, most inhabiting the bottoms of temporary pools of freshwater and feeding on small particles, either directly from the water or from the sediment they stir up. Fossil concostracans, which are known back to the Devonian, can be very abundant in some freshwater deposits and have been used for stratigraphical purposes to subdivide sequences such as the Mesozoic of the Sichuan Basin, China.

Ostracods are a geologically important group of crustaceans ranging from the Cambrian to the present day where they number about 2000 species. Like conchostracans, ostracods have bodies enclosed between the two valves of a carapace, but unlike conchostracans this is generally calcified – hence ostracods have an excellent fossil record – and tends to be more rotund, often bean-shaped. Different species inhabit a wide range of environments, from forest soils to the ocean floor, feeding in a variety of different ways and exhibiting different degrees of mobility. Most fossil ostracods are less than 1 mm (0.04 inches) in size and therefore fall within the realm of microfossils. However, a few species reach a size of 3 cm (1 inch).

The final group of crustaceans routinely fossilised are the barnacles, or cirripedes, with some 1000 living species. As a consequence of their attached mode of life, barnacles have highly modified bodies setting them apart from the other crustaceans. Barnacle larvae are of the typical crustacean type but further development is highly unusual, the larva becoming attached to a hard surface by its antennae, discarding the carapace and secreting a series of protective calcareous plates around

**Above** A Cretaceous oyster from Georgia, USA peppered with the tiny slit-like borings made by acrothoracican barnacles, each about 1 mm in length.

the body. The adult barnacle (see colour fig. 27) is almost like a shrimp that sits on its back and wafts its legs in the air to drive water laden with plankton towards the mouth. Because of the resistant calcareous plates, barnacles have a good fossil record. The earliest examples date from the Silurian and belong to the so-called 'goose-necked' barnacles, in which a thick fleshy stalk attaches the animal to a hard surface. Driftwood in the sea today is not uncommonly colonised by goose-necked barnacles that hang from the underside. Most fossil examples of goose-necked barnacles consist of scattered plates left after decomposition of the stalk and other soft parts; articulated specimens are far less common.

More familiar to beachcombers are the acorn barnacles. These lack stalks and instead are cemented directly to hard surfaces such as shells and rocks. They are among the relatively few marine animals that have adapted successfully to the rigours of life in intertidal environments where recession of the tide brings the constant danger of desiccation. In the case of acorn barnacles, four opercular plates at the top of the cone-shaped shell can close together tightly, sealing the animal until the water returns. Acorn barnacles are surprisingly late entrants into the fossil record, not appearing until the Cretaceous. Even then they remained rare until the late Eocene, about 40 million years ago, when the group of acorn barnacles – balanomorphs – dominating in today's seas first appeared. From this time onwards, acorn barnacles became common, their fossilised plates occasionally being present in rock-forming quantities. Barnacle plate limestones, for example, occur in the Pliocene of New Zealand. Fossil acorn barnacles are most commonly encountered attached to oysters and other mollusc shells. The most completely preserved examples retain the opercular plates, but these usually drop out and it is more usual to find specimens preserving only the four to eight compartmental plates forming the sides of the conical shell, plus the basal plate that is cemented to the substrate.

When not accumulating evidence to support his theory of evolution through Natural Selection, Charles Darwin devoted much of his energy to the study of fossil and living barnacles. He became fascinated by these crustaceans after discovering in Chile an unusual type of tiny barnacle which, instead of making a calcareous shell, lives in holes it bores into mollusc shells and similar hard substrates. Borings made by such acrothoracican barnacles can be found back to the Devonian. These trace fossils have a characteristic pouch-like form, with a slit-shaped or oval opening through which the feeding appendages of the living animal once protruded.

## ARCHAEONISCUS: JURASSIC–CRETACEOUS OF EUROPE

The body of this sea-slater is oval in outline shape and strongly compressed dorso-ventrally. The cephalon has a rounded rectangular shape with eyes present near the outer edges. Segmentation is conspicuous in the rest of the body, which consists of an anterior region called the perion and a posterior pleotelson of semicircular shape.

The isopod *Archaeoniscus* is found in some abundance in the Purbeck Beds of southern England where it seems to have been able to cope with the higher than normal salinities pertaining during deposition.

**Above** *Balanus*, a large (9.6 cm high) acorn barnacle of Pliocene age collected in Suffolk, England.

## BALANUS: EOCENE–RECENT, WORLDWIDE

The conical shell of this acorn barnacle consists of six fused plates that have hollow walls. A basal plate is cemented to the substratum, and the opening in the top of the conical shell can be closed by four triangular plates comprising two tergal and two larger scutal plates.

*Balanus* is extremely abundant along modern rocky shorelines. In the UK, individuals of *Balanus crenatus* live for 1–2 years, grow at a rate of 4–5 mm (0.2 inches) per month to a maximum size of 25 mm (1 inch), and produce larvae that swim in the sea for more than a

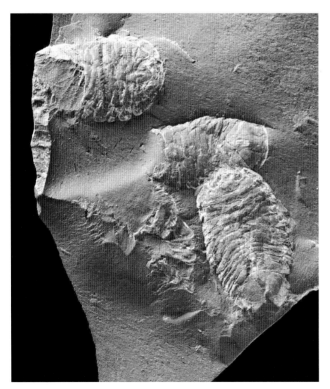

**Above** Three individuals of the Cretaceous isopod *Archaeoniscus* from Wiltshire, England. The largest measures 1.5 cm.

month before settling and metamorphosing into the permanently fixed adult. Larger species of *Balanus* may have longer life spans, *B. cariosus* reaching 100 mm (4 inches) in diameter and living for up to 15 years in California. Fossil specimens of *Balanus* are usually found without the opercular plates. Sometimes only the basal plate is preserved on shell and rock substrates, looking at first like the cemented valve of an oyster.

### *HOPLOPARIA*: CRETACEOUS–MIOCENE, WORLDWIDE

*Hoploparia* is a lobster with an elongate, slightly compressed body. Fine granules cover the carapace (thorax). Projecting forward of the head, the rostrum is long and thin. Antennae are also long. The elongate abdomen curves beneath the carapace and ends in a rounded telson. There are six pairs of thoracic appendages, the first pair used for feeding, the second pair with claws (chelae) of which that on the left side is long and narrow whereas that on the right side is stout and toothed.

**Above** Compressed specimen of the crab *Portunus*, from the Oligocene of the Caucasus, showing a swimming leg with its paddle-shaped dactylus (lower right).

Unusually for a lobster, this genus has unequal-sized pincers, a feature more commonly associated with many true crabs and hermit crabs.

### *PORTUNUS*: EOCENE–RECENT, WORLDWIDE

This crab has a very broad and flat carapace with a granular surface, three to six teeth at the front end and usually nine teeth along the forward-facing lateral margins. The eye stalk is elongate and the claws are long. A paddle-shaped final segment (dactylus) occurs on the fifth leg, and all of the other segments of this leg are flattened.

*Portunus* is a typical swimming crab as indicated by the paddle-shaped final segment of hindmost leg.

### *STRAMENTUM*: CRETACEOUS OF NORTH AMERICA, EUROPE AND ASIA

The goose-necked barnacle *Stramentum* has a capitulum atop a stalk (peduncle) which would have been basally attached to a hard or firm surface during life. The barnacle is oval or lens-shaped, the capitulum smaller than the stalk and consisting of nine plates arranged in a single whorl. Eight vertical rows of overlapping, imbricated plates make up the stalk.

**Above** An exquisite specimen of the Cretaceous lobster *Hoploparia*, with a body measuring over 12 cm in length, from Dorset, England.

**Above** Cretaceous goose-necked barnacle *Stramentum* from the Chalk of Kent, England, about 2 cm wide.

Most lepadomorph barnacle fossils like *Stramentum* comprise disarticulated and scattered plates, but intact specimens are occasionally found signifying very swift burial. Some evidently lived attached to ammonites or inoceramid bivalves.

## MYRIAPODS

Millipedes (Diplopoda) and centipedes (Chilopoda) belong to a major division of arthropods called the Myriapoda. A distinct head, with antennae, is followed by a long body comprising numerous segments. There is no clear distinction between a thorax and an abdomen. Most of the segments have one pair of walking appendages that are unbranched (uniramous), contrasting with the biramous appendages of trilobites for example. All modern myriapods are terrestrial animals, using tracheae to respire in the air. Although fossil myriapods are found dating back to the Silurian and were the first animal group known to have colonised the land, they are never very numerous, and fewer than 100 fossil genera are known. The recently discovered

**Above** Measuring 7.1 cm long, this Carboniferous fossil represents only part of a leg of the giant millepede *Arthropleura*.

*Pneumodesmus newmani* from mid-Silurian deposits of Scotland preserves openings to the trachea called spiracles, showing that it was an air-breathing, terrestrial millipede.

Without doubt the most spectacular fossil myriapod is *Arthropleura*. This millipede-like Carboniferous animal has been claimed to be the largest terrestrial arthropod ever to have lived, attaining a length of 2 m (6.5 feet) or more. Specimens are invariably fragmented

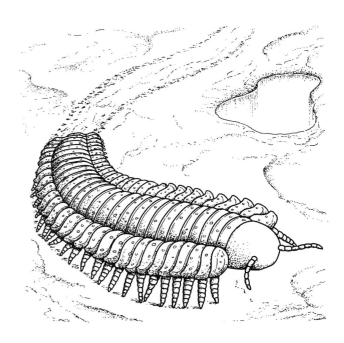

**Above** Reconstruction of the enormous millipede *Arthropleura* which may have reached at least 2 metres in length.

but the complete animal is estimated to have had about 30 pairs of legs. Some spectacular examples have been described of trails – *Diplichnites* – left by individuals of *Arthropleura* (or a related genus) walking across damp sediments. Despite its gigantic size, *Arthropleura* probably fed on vegetation and/or detritus and was not a carnivore.

## INSECTS

It has been estimated that at least one in every three animal species alive today is an insect. There are thought to be far in excess of 1 million species of living insects, including some 300,000 named species of beetles alone with many more yet to be formally described. Being diverse and abundant as well as overwhelmingly terrestrial in ecology, insects are the most familiar of all invertebrates to humans, many loathed (e.g. cockroaches, wasps, flies, ants and termites) but others loved (e.g. butterflies and bees).

Insects have a body of three parts: the head with six segments, thorax with three segments and abdomen with 11 or fewer segments. Two compound eyes are present on the head, which also has sensory antennae and chewing mandibles. Each of the three thoracic segments has a pair of legs and the second and third thoracic segment also commonly have a pair of wings. There are neither legs nor wings on the abdominal segments. The animal therefore has six legs in total and four wings, although in some insects (e.g. flies) one pair of wings is absent and in others (e.g. fleas) there are no wings at all. Insect appendages are unbranched (uniramous), like those of myriapods, and respiration is accomplished similarly using trachea.

The fossil record of insects is very patchy. Many species are known from just a single specimen, and a high proportion of fossil insects come from exceptional fossil deposits (*Lagerstätten*) that are sporadically distributed in time and space. Nonetheless about 40,000 species of insects have been described as fossils, with many more awaiting description. Paramount among

insect-rich deposits are ambers in which complete external preservation of insects is routine. Amber is the fossilised resin of a few particular kinds of trees. Oozing out of the bark, this resin had the ability to trap and encapsulate insects, as well as other animals, protecting them from the normal processes of organic decay as it hardened into transparent, yellow or orange coloured amber. The chemical process of 'amberisation' could take up to 10 million years. During this time it was common for amber initially buried in the soil to be washed out by rivers and redeposited in the sea. Although the oldest amber comes from the Carboniferous, the great majority of amber deposits are of Cretaceous or Cenozoic age. They provide priceless 'windows' on the insects and other small animals living at the time in the forests where amber-producing trees grew. Elsewhere in the fossil record, insects can be found in fine-grained sedimentary rocks, such as clays and silts deposited in freshwater lakes and sluggish rivers. Unlike the insects in amber, these fossils generally comprise only fragments, particularly of wings or wing cases, although more complete examples can be found, such as the dragonflies of the famous Jurassic Solnhofen Limestone of Bavaria (see colour fig. 28).

It is impossible to do justice in a short space to the multiplicity of insects described from the fossil record, let alone to review the immense diversity of recent insects. The most basic categorisation of insects is into a primitive group without wings, called the Apterygota, and the winged Pterygota. Surviving apterygotes include the springtails and silverfish. They are relatively rare, comprising less than 1 per cent of all insect species. Pterygotes are divided into those with wings that cannot be folded, which are called the Palaeoptera, and a larger, more advanced group, the Neoptera, capable of folding their wings close to the body. Mayflies, dragonflies and damselflies are all palaeopteran insects. Neopterans are further subdivided into forms having immature stages (nymphs) that resemble small adults, so-called 'incomplete metamorphosis', and others exhibiting complete metamorphosis in which the egg hatches into a larva (e.g. a caterpillar), the larva produces a pupa (e.g. a chrysalis) which in turn metamorphoses into the adult form (e.g. a butterfly). Insects having incomplete metamorphosis include cockroaches, termites, locusts, earwigs, stick insects, bugs and lice. Complete metamorphosis occurs in, for example, butterflies, lacewings, fleas, flies, ants, wasps, bees and beetles.

In spite of its imperfection, the fossil record holds a lot of useful information about the times of origination of insect groups that are alive today, and also, of course, provides the only evidence of groups that are now totally extinct. Primitive wingless insects – the apterygotes – appear to have undergone an initial diversification during the Devonian, possibly even the Silurian. Unfortunately, however, relatively few fossil insects of this age are known and there is a great need for further prospecting. The oldest known fossil insect is currently *Rhyniognatha hirsti* from the early Devonian Rhynie Chert of Scotland. However, this species, preserved in sinter from an ancient hot water spring active between 400 and 412 million years ago, exhibits some advanced characteristics implying that there are more primitive, older insects still to be discovered.

Fossil insects with preserved wings (Pterygota) first occur in the mid Carboniferous. The evolution of wings was accompanied by an increase in maximum body size. A remarkable dragonfly called *Meganeura* with a wingspan approaching 70 cm (27 inches) has been described from the late Carboniferous. This inhabitant of the Coal Measure forests is one of the largest insects ever to have lived. The huge size of *Meganeura* has led to speculations about the composition of the atmosphere at the time, the

**Above** Small concretia, 4 cm high from the Carboniferous of the English Midlands, containing the cockroach *Aphthoroblattina*.

powered flight of such a large insect perhaps demanding an atmosphere containing higher levels of oxygen than that of the present day. Unfortunately, the fragmentary insect fossil record sheds little light on the origin of flight, as the oldest winged insects already have fully-formed wings. It has been hypothesised that wings originated from flaps that helped animals to land right-side-up when jumping or blown by the wind. Enlargement of these flaps into proto-wings may have occurred as an adaptation for effective parachuting or gliding from tall plants.

Known only from the wings, the body of the Carboniferous insect *Brodia* has yet to be discovered. The wings of some examples of this fossil are remarkable in preserving their original pigmentation: three or four dark bands alternate with paler coloured bands along the length of the wing. Patterns of venation on the wings of fossil insects such as *Brodia* are of immense importance to palaeontologists when identifying species and interpreting which major group of insects they belong to.

The ability to fold the wings, of which dragonflies are incapable, had evolved in neopteran insects by the late Carboniferous. Cockroaches, for example, can be locally abundant in sedimentary rocks of this age. Like the dragonflies, Carboniferous cockroaches could attain a very large size: an example 9 cm (3.5 inches) long has been found recently in an Ohio coal mine. An artificial grouping of insects (protorthopterans) also occur in some abundance in the late Carboniferous. It is thought that protorthopterans encompassed a wide range of feeding types, with some species being herbivores but others scavengers or predators. They declined during the Permian and very

**Above** Pigmented wing of the Carboniferous insect *Brodia* measuring over 4 cm in length.

**Above** Delicately preserved crane fly *Tipula*, with a body length of 2.9 cm, from the Florissant Fossil Beds of Colorado.

few survived the mass extinction at the end of the period, which was calamatous for insects in general.

Several important extant groups of insects can be traced back to the Permian, such as the book-lice, stone-flies, bugs and beetles. Further groups appeared in the Mesozoic, including moths, ants and bees, with butterflies appearing somewhat later, in the Eocene. Termites make their debut in the Cretaceous fossil record. They were probably derived from cockroaches that adopted a social lifestyle of living in colonies. Modern examples of termite colonies can contain up to 5 million individual insects. Several groups of insects, butterflies and bees especially, undoubtedly diversified in concert with flowering plants (angiosperms) during the late Cretaceous and Cenozoic. A great many insect species are specialist feeders on one or a small number of angiosperm species, and as new species of angiosperms appeared so new species of insects evolved to feed or pollinate them, a process known as coevolution. Unlike many other vertebrate and invertebrate groups, insects show only minor indications of suffering a mass extinction at the end of the Cretaceous.

In addition to body fossils of insects it is also possible to find their trace fossils. For example, the cases constructed by caddis-flies to house their larvae that dwell in freshwater environments are occasionally fossilised in Cretaceous and younger deposits. Termite nests have been recorded from several Cenozoic fossil soil deposits. However, perhaps the most common insect trace fossils are found associated with plant fossils. These include borings made by beetles in fossil wood, and sinuous traces made by leaf-mining insects feeding on leaves.

*TIPULA*: PALEOCENE–RECENT, WORLDWIDE
This crane fly has a slender body and very long legs. The antennae are thread-like, with many distinct segments, and the wings are long and narrow.

Adult crane flies are abundant today in moist, wooded environments. Their larvae are worm-like, most being aquatic and often feeding on leaf mould. 20 species of *Tipula* have been described from the famous Florissant Fossil Beds of Colorado, USA. These Eocene deposits, which also contain many fossil plants, were formed in an ancient freshwater lake.

# 6

# Spiny-skinned animals

SEA-URCHINS (echinoids), sea-lilies (crinoids), starfish (asteroids), brittle-stars (ophiuroids) and sea cucumbers (holothurians) are all groups of echinoderms that live today and are also known from the fossil record. The name 'echinoderm' means 'spiny skin', referring to the spinose bodies of sea-urchins in particular. Several different kinds of extinct echinoderms occur in the Palaeozoic, providing a record of the early evolution of this important phylum of invertebrates. Echinoderms have three basic characteristics: they have high magnesium calcite skeletons, bodies with pentameral symmetry (a five-rayed arrangement of the body), and a water vascular system.

The skeletons of echinoderms are formed of many plates, each plate being a single crystal of calcite. Under high magnification the crystalline plates of modern echinoderms have a spongy texture referred to as 'stereom'. In life, the pores of stereom are filled with soft tissue including collagen ligaments and muscles which hold the plates together. When a modern echinoderm is damaged, the calcite is more likely to break across the plate than along the boundaries because of the great strength and resilience of the collagen and muscular tissues. Fossils can also show this feature.

Fossil echinoderm plates are generally secondarily mineralized, with the 'holes' in the stereom becoming infilled by calcite that develops as an overgrowth on the skeleton. Because the secondary calcite has the same crystallographic orientation as the original calcite of the plate, plates of fossil echinoderms are still single crystals. A characteristic of fossil echinoderm plates is that they break along calcite cleavage planes. The flat or stepped fracture surfaces are lustrous and provide a useful means of identifying unknown fossils as echinoderms.

Individual plates may interlock or may be held together only by soft tissues during life. After death and the decay of the soft tissues, the plates often separate, and many fossil echinoderms consist of dissociated plates and spines. However, intact skeletons can be preserved if burial is rapid and precedes the onset of soft tissue decay. Although all echinoderms have skeletons composed of individual plates of calcite, in holothurians the skeleton is often no more than tiny spicules embedded in the flesh of the animal. These generally become scattered after the animal dies.

The pentameral symmetry of the echinoderms is not perfectly radial. The presence of a madreporite in some (e.g. echinoids, starfishes), or anal tubes in others (e.g. crinoids) prevents perfect radial symmetry. The five-

**Left** *Dimerocrinus*, a small stemmed crinoid ('sea-lily') of Silurian age. The long stem supports a 1.8 cm high crown mostly consisting of the arms employed in feeding.

rayed arrangement may also become bilaterally symmetrical as in some 'irregular' sea-urchins. Several species of starfish routinely have more than five arms, whereas sea-urchins are occasionally found with more or fewer than five-rays as a result of genetic abnormalities or mechanical damage to the living animal, though the basic skeletal plan is still five-rayed. The spiral arrangement of helicoplacoids is another anomaly.

Echinoderms have a water vascular system controlling the operation of the extensible tube-feet, which are especially noticeable on the outsides of the bodies of echinoids and starfish. The function of the tube-feet varies according to the kind of echinoderm and includes locomotion of the animal as well as movement of sediment particles, respiration, feeding, burrow construction, and chemoreception.

The echinoderms are divided into two subphyla which distinguish their essential mode of life. The Pelmatozoa are those echinoderms with a calyx and which are fixed to a substratum either directly by cementation, by the development of a holdfast, or by cirri or radices of the stem. The loss of the stem in some crinoids is a secondary adaptation to a free living life-style. Pelmatozoans have their mouths directed away from the substratum and sediment and rely on radially arranged arms together with tube-feet to capture and gather food particles. Extinct forms include the cystoids, blastoids and eocrinoids. Only the crinoids survive today as representatives of the group. Most species of crinoids throughout the long history of the group were tethered to a substratum by a stem, although free-living dominate at the present day. Crinoids are the only group of echinoderms living today that include species attached by a stem; all other types of echinoderms which had stems are now extinct.

The Eleutherozoa are the echinoderms that are free-living and include all the other modern day extant classes: sea-urchins (echinoids), sea-stars or starfish (asteroids), brittle-stars (ophiuroids) and sea cucumbers (holothurians). The edrioasteroids of the Palaeozoic belong to this group but were cemented directly to a substratum and had no stem. Movement around the sea floor is achieved by means of the tube-feet, spines, and muscular arms, and sometimes all of these. The mouth is directed towards the substratum so that they can ingest sediment or graze. Starfish and brittle-stars can also be active predators, and sea-urchins sometimes feed on moribund animals or may on occasions capture small prey.

Several major classes of now extinct echinoderms lived during the Palaeozoic. These include the cystoids, blastoids, eocrinoids, edrioasteroids, helicoplacoids, ophiocistioids and a few more minor groups. Not all extinct echinoderms have an obvious pentameral symmetry, or indeed any trace of pentameral symmetry, but all do possess skeletons composed of calcitic stereom. Animals closely-related to, but not considered to be true echinoderms by some palaeontologists, also possess stereom. These are the carpoids which probably shared a common ancestry with the echinoderms.

Some of the extinct groups of echinoderms that lived in the Palaeozoic were rock formers – cystoids occasionally formed limestones, such as the Ordovician *Echinosphaerites* Limestone of Sweden, as did blastoids in the Carboniferous Limestone of Britain. However, the extant crinoids are more often present in rock-forming quantities, as is evident by the profusion of crinoidal limestones in the geological record.

Modern echinoderms exclusively inhabit marine environments – they are never found living in freshwater habitats. The same is believed to have been true in the geological past and therefore fossil echinoderms serve as reliable indicators of rocks deposited in marine conditions.

## ECHINOIDS

Popularly known as sea-urchins, echinoids can be divided into main two types. Radially symmetrical forms, the so-called 'regular' echinoids, have a more or less globular shape, whereas bilaterally symmetrical forms, the 'irregular' echinoids, are more variable. The upper side of both types is referred to as the apical surface and the underside as the oral surface. The mouth of an echinoid is contained within an area called the peristome, and the anus within the periproct. The position of the peristome and the periproct varies according to the type of echinoid. Both the mouth and anus have protective coverings of small plates. The peristome of regular echinoids is relatively larger than that of irregular echinoids.

On the apical surface of regular echinoids is a structure called the apical disc (or 'disc' for short) or apical system, which has five ocular plates and five genital plates. The ocular plates have a pore through which passes a sensory tentacle. The genital plates are also perforated, four of them by a single pore and one of them by several pores. This porous plate is called the madreporite and is connected to the water vascular system. The oculars are radial in position and the ambulacra extend from them. The genital plates are interradial and the interambulacra extend from these.

Irregular echinoids also have an apical system but they are rather differently arranged. The periproct is no longer included within the structure but is found elsewhere on the test, ranging in position from the apical to the oral surface. Ocular plates and genital plates are present and these vary in number and position, and are important in classification.

In most urchins, both regular and irregular, the eggs and spermatozoa are similar in size and adults do not show sexual dimorphism. Eggs and sperm are released into the water where fertilisation occurs, from which a free swimming larva develops. However, in some urchins the females produce larger eggs, though fewer in number, which remain either amongst the spines or in special depressed areas of the test known as 'brood pouches'. The larger eggs may be the result of internal fertilisation whereby sperm are taken in from the water. There is no free swimming larval stage, but instead the new urchin develops directly. In the two sexes of brooding urchins, the females and males show distinct dimorphism, with female genital pores larger than those of the males. Most of the brooding urchins are nowadays found in colder waters, though some are found in tropical areas. Evidence of sexual dimorphism in fossils is limited to the more obvious structural morphology.

In all post-Palaeozoic echinoids there are five thinner rays called ambulacra and five thicker ones called interambulacra. Each ray is composed of a pair of columns of interlocked calcite plates. Those of the interambulacra are larger than those of the ambulacra. The ambulacral plates are perforated for the passage of the tube feet, and are referred to as the pores or pore-pairs. The plates may be simple or they may be compound, with a plate made of two or more components. The compounding is very important in the classification of echinoids, with types of compounding being characteristic of well-defined groups. New plates are added adjacent to the apex of the test and the first-formed plates surround the mouth. The ambulacra of irregular echinoids vary considerably in outline and extent, from simple narrow columns flush with the general surface of the test, deeply sunken, or broadly inflated and petal-shaped. The general trend in irregular echinoids is towards increasing bilateral symmetry.

Echinoids use tube-feet to assist with movement, burrowing, sensory activities and respiration, and these pass out of the body by means of paired pores in the ambulacra. In some Palaeozoic echinoids the ambulacra are wider than the interambulacra, and both may have many more columns of plates than just the modern two columns per ray.

Tubercles are present over the surface of the test, which bear moveable spines (or 'radioles'), also made of calcite; these cover the test (the 'shell') of the urchins. The primary spines vary in length from short and robust to long and delicate in regular urchins, and fine and hair-like on irregular urchins. They are attached to the tubercles by ball-and-socket joints and muscles. The spines are for protection against predators, for support and movement over the surface of the sea bed, or they may assist in burrowing or excavating. Secondary tubercles bear smaller spines which act as protection for the soft musculature which controls the movement of the primary spines. Tiny specialised appendages called pedicellariae, some of them toxic, are also attached to other, minute tubercles and these serve several purposes, including defence against parasites and small predators, cleaning by removal of foreign particles, and feeding, whereby food particles or small animals are captured and passed to the main feeding areas.

The mode of life of fossil sea-urchins can be inferred by the kind of spines they had. In regular urchins, coarse and robust spines are useful for defence (see colour fig. 31), and long and fine spines are also useful for supporting the urchin on a softer substratum. Spines may be smooth or they may have some kind of ornament, often in the form of longitudinal striations, or spinules – tiny thorns along the shaft of the spine. The fossil regular urchin *Tylocidaris*, from the Cretaceous Chalk of southern England, has relatively massive club-shaped spines which if they were solid would take considerable force to move, but they are spongy on the inside. Their purpose was probably both for defence and for support on the soft Chalk sea bed.

Perhaps the most beautiful forms of spines are those of the diadematids. These spines are called 'verticillate', characterised by rings of thorns that appear to be set one inside another to produce a very brittle and delicate spine (the spine is still a single crystal of calcite even then). The cores are hollow or loosely spongy. Diadematids can respond to changes in light and will bunch clusters of spines together and move them towards a potential threat in the water above them. These urchins are often the cause of much distress and pain to swimmers who come into contact with them; as they are delicate, they snap off very easily and embed themselves in the flesh of the victim. Apart from treating the wound with antiseptic and pain killers, the best way to get rid of the spines is to break them up in the wound and let the body take care of the remains!

In irregular urchins, the fine hair-like spines (see colour fig. 29) densely covering the test of a burrower keep the sediment through which it burrows away from the surface of the test. Functional adaptations of certain spines of irregular urchins allow the animal to burrow, whilst the solid spines of some regular urchins have become spongy and club shaped.

Many regular echinoids have a complicated jaw mechanism called the 'Aristotle's Lantern', also made of calcite, which may rasp or crush food. A lantern has 40 individual components, and the complete jaws are operated by 60 muscles; some of the muscles move the teeth up and down, others keep the components in place relative to one another, and some attach the whole lantern to the test or the perignathic girdle (a special attachment structure around the inner surface of the peristome) if present. The lantern is thus suspended within the test and is allowed considerable movement.

Other echinoids have lost the jaws (except in juveniles) and gather food particles by means of specialised tube-feet and spines around the mouth.

Food gathering by sea-urchins nowadays is by a variety of methods, including active grazing and relatively passive sediment feeding, but the early forms of the Palaeozoic were probably detritus-grazing feeders; the preserved jaws of some of them are very similar to those of modern regular urchins.

Regular urchins live mostly on the surface of the sea bed, or in protected rocky areas, sometimes excavating holes in rocks. They are grazers or scavengers, and some may even feed on moribund animals. Groove-like grazing traces called *Gnathichnus* can sometimes be found on fragments of fossil mollusc shells, and are five-rayed (pentastellate), showing the rasping scratches made by the teeth. The presence of this trace fossil provides palaeontologists with clear evidence that feeding by grazing on hard surfaces had evolved by the Triassic, if not before.

**Above** *Gnathichnus*, the trace fossil made by a grazing echinoid on the surface of a British Jurassic oyster. Field of view is 3.8 cm.

Regular urchins feed on a variety of food, including seaweeds and plant debris either attached or drifting, detritus on the bottom, encrusters and boring organisms, and sessile animals. They will also feed on more active animals if they come into range. The jaws are sufficiently strong to crush mollusc shell-fragments despite being made of calcite, and they will also feed on moribund animals. They can also absorb chemical foods directly into the epithelia of the soft tissues, though this is not a major source of nutrition. The urchins may also show a slight preference for some kinds of food over others, and can detect potential food at a distance. This is done by the use of specialised chemoreceptors, and sampled for suitability by the tube feet of the peristome. Food is manipulated into the mouth by means of spines around the peristome and by the tube feet. Some regular urchins which bore into rocks and live within the holes, sometimes become too large to come out again. Grabbing hold of passing food material – plant debris, small animals and so on – then becomes very important; they achieve this by extending the aboral tube feet which are equipped with suckers to catch such material.

Irregular urchins are partly or entirely infaunal – living within the sediment in burrows. They are deposit feeders or detritivores, and consume sediment. Echinoids which live within the sediment but do not produce burrows transfer individual particles of the surrounding substrate over their upper surface. Whereas burrowers can live in fine sediments, these echinoids prefer coarser sediments with a more open sediment pore space.

Apart from one group of modern urchins (clypeasteroids) which have strong jaws used for crushing rather than scraping food from surfaces, most irregular urchins ingest sediment and extract food particles from it because they do not have a strong jaw apparatus. The feeding regimes are various and include living on or just within sands and gravels which they ingest to extract the organic matter, e.g. the sand-dollar *Dendraster* which can live either horizontally or slightly inclined into a mild water current. Other irregulars burrow into the sediment and construct elaborate funnels to the surface. Burrowing echinoids produce sanitary tubes and respiratory funnels by means of their spines and tube-feet. Mucus is produced by the tube-feet and it is transferred by them, and by the spines, to the tunnel walls to strengthen them. Minute ciliated spines called clavulae are attached to tiny tubercles which form the fascioles of some irregular echinoids. These create water currents in and out for the animal living in burrows, and thereby assist in feeding and other functions. There are several kinds of fascioles, each type named for its location on the test. Sediment or food particles within the sediment are extracted by oral tube-feet, or removed from mucus strands.

Food particles are collected by prehensile tube feet extended towards the surface, with some also using mucous strings to trap and transport food particles to the mouth in a sort of conveyer belt. Spines over the surface of the test prevents the coarser sediment particles from reaching it but allows the finer food-rich particles to do so. From there the food is transported by mucous strings to the mouth, and in some types of urchins by specialised food grooves e.g. *Dendraster*. Fossil burrows and feeding traces of irregular urchins are known, as well as the scratch marks made by the special enlarged spines

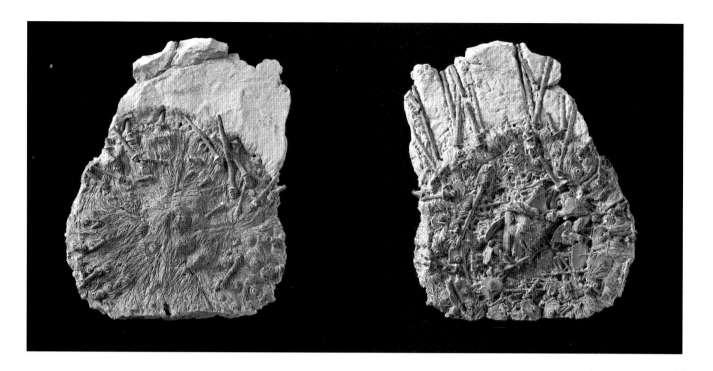

**Above** *Archaeocidaris whatleyensis*, 9 cm across, from the Carboniferous Limestone of Somerset, England; apical (left) and oral (right) views.

during movement in the burrow. These trace fossils are called *Subphyllocorda* and *Scolicia*, both 'ichnogenera', which describe the trace fossil rather than the animal that made it.

Waste removal is equally important to the echinoids. Regular echinoids rely on being in the open water and perhaps on current to remove waste products. This option is not available to infaunal burrowers and other less active urchins. The periproct has, however, been moved away from the oral and respiratory areas of the test to an opposite end where waste can effectively be left behind when the animal moves through the sediment or within its burrow.

Echinoids occur worldwide in rocks of the Ordovician to Recent. They are relatively uncommon in Palaeozoic rocks but become ever more common through the Mesozoic and Cenozoic. Modern species, of which almost 1000 are known, occur from shallow waters down to abyssal depths.

## *ARCHAEOCIDARIS:* CARBONIFEROUS OF NORTH AMERICA, EUROPE, ASIA, AUSTRALASIA

This regular echinoid, up to 10 cm in diameter, has a low profile. Test plating is imbricated which enabled the test to be flexible during life. The apical disc is monocyclic (arranged in one circle) and has five ocular plates and five genital plates, including the madreporite. The ambulacra are composed of two columns of small and 'simple' polygonal plates, with flanges that extend beneath adjacent interambulacral plates; pore-pairs are uniserial (in a single column) and the tubercles minute. The interambulacra each have four columns of plates which are large and hexagonal. The primary tubercles are large, perforate and non-crenulate, and occupy a major portion of each plate. The peristome is small and internally without auricles (attachment structures for the jaws). The Lantern of Aristotle (jaw apparatus, usually referred to as the 'lantern') is low angled. The spines are very long, with coarse spinules (thorns), sometimes hollow and are often preserved in a crushed condition.

*Archaeocidaris* is one of the most advanced Palaeozoic echinoids. It belongs to a group that evolved long spines, presumably for defence, but perhaps also for support on a soft substrate, like some modern and later fossil forms. Although it had four columns of interambulacral plates, in other respects it resembled the later *Miocidaris* of the Mesozoic, the first of the true cidaroids.

## *CLYPEASTER:* EOCENE-RECENT, WORLDWIDE

Measuring up to 12 cm in length, this irregular echinoid has a robust, rigidly plated test with a hemispherical, domed or subconical profile, and ovoid to lobed pentagonal outline. The oral surface is flattened. The margins are rounded or flattened, sometimes inflated. The apical disc is central and pentagonal or star-shaped, with four gonopores and with the madreporite greatly expanded. The apical ambulacra (those of the upper surface) are all petaloid, wide and may be inflated in the interpore zone – the central portion of the petal between the two columns of pores. The adoral ends of the petals may be closed or open. The pore pairs are unequal in size, the outer ones being very elongate but the inner ones small. Pores adoral to the ends of the petals are tiny and single. The interambulacra are narrower than the ambulacra, and on the oral surface they are not differentiated. The peristome is central, circular and is sunk deep into the test. There are no buccal notches, but there are five fairly shallow, straight and unbranching food grooves from the mouth almost to the margin of the test. The lantern is robust. The periproct is small, circular and situated inframarginally, i.e. close to the posterior margin but still just within the oral surface. The test is much strengthened by a robust internal system of buttresses and pillars which connect the upper and lower surfaces. These occupy a considerable amount of the internal space. Suitable holes within these structures permit the normal configuration of soft parts to be developed. The tubercles of the oral and apical surfaces

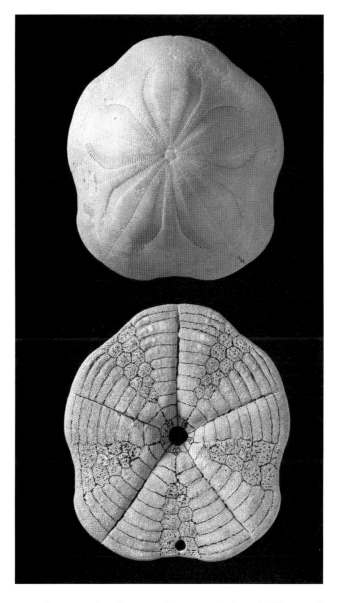

**Above** *Clypeaster altus,* 13 cm anterior to posterior, from the Miocene of Malta; apical (upper fig.) and oral (lower fig.) views.

are very small, uniform and close packed, and the radioles are small, short and simple.

Most modern species of *Clypeaster* inhabit shallow waters, but a few live at 200 m or more. It is a deposit feeder living on or partially buried in coarse sediments. There are some 400 species assigned to this genus, but some may need to be reassigned to other genera as knowledge increases.

## DENDRASTER: MIOCENE-RECENT, NORTH AMERICA

*Dendraster* is an irregular echinoid up to 8 cm in length, with a rigid, almost circular test, usually wider than long and flattened in profile, somewhat biscuit-shaped. The apical disc is small, with four gonopores and situated posterior of centre. The ambulacra are petaloid with the petals flush with the surface and ending abruptly. The anterior petal is slightly longer than the others. The peristome is small, central, circular, and has no buccal notches. The interambulacra on the oral surface are about as wide as the ambulacra. The periproct is near the posterior margin (inframarginal) and is small and circular. The tubercles are uniformly small and dense. There is a complex branching system of food grooves on the oral surface which radiate from the mouth, and these extend aborally around the posterior margin but not to the anterior margin. The lantern is completely internal, with the perignathic girdle composed of five interambulacral projections, designated 1–5. The spines are small and simple.

Dendraster is the only suspension feeding echinoid. The anterior part of the test is pushed into the sediment, thus raising the posterior part into the water current to capture plankton and other food particles. This echinoid is also a deposit feeder and lives in shallow water in large numbers. *Dendraster* is one of several echinoids referred to as 'sand-dollars'. It is immediately recognisable by its flattened appearance and its eccentric apical system.

## ECHINOCARDIUM: OLIGOCENE-RECENT, WORLDWIDE

*Echinocardium* is a moderately inflated, heart-shaped, irregular echinoid with a rigid but very thin and fragile test up to 50 mm long. It has a small, central apical disc with four gonopores. The anterior ambulacrum forms a

**Above** *Dendraster excentricus*, 7.9 cm in diameter, from the Pliocene of California, apical view.

shallow groove (anteal sulcus) from disc to peristome, and is not petaloid. The paired ambulacra are petaloid with each column closing at the adoral ends and widening adapically to form a broad star-shape. The pores are paired within the petaloid region but are single adorally. There is an internal fasciole, which encloses the apical disc and part of the anterior test, and a subanal fasciole and an incomplete circum-anal fasciole which extends adapically and towards the anterior as two branches in an extended, nearly closed U-shape, almost reaching the internal fasciole. The tubercles of the apical surface are small and closely packed, except adjacent to the anteal sulcus where they are rather larger. Here, tufts of spines are present in life. On the oral surface, the plastron has paired and symmetrical sternal and succeeding plates. The tubercles are larger than those of

**Above** *Echinocardium orthonotum*, 4.5 cm across, from the Miocene of Maryland; apical (left) and oral (right) views.

the apical surface and differentiated into defined areas. The labrum is short and wide and overlies a D-shaped peristome. There is no lantern or perignathic girdle. The periproct is situated high up on the flattened posterior of the test and has tufts of spines below it.

*Echinocardium* excavates a burrow in the sand up to 12 cm below the surface of the sea bed. A narrow funnel to the surface enables it to draw water into the burrow for feeding and respiration, and a second horizontal funnel is used for drainage. *Echinocardium* is a deposit feeder, utilising small particles. Broken, damaged tests of *Echinocardium* are often washed ashore on sandy beaches after storms, having been scoured out of their burrows by wave action.

## *MELONECHINUS*: CARBONIFEROUS OF EUROPE, ASIA AND NORTH AMERICA

*Melonechinus* is a regular echinoid with a globular test, up to 10 cm in diameter, composed of thick polygonal plates. The apical disc is small and has five ocular plates

and five genital plates which are each perforated by several gonopores. The ambulacra have many columns of small plates each bearing a simple pore-pair. The two central columns are widest. There are bands of small plates developed adradially. Up to five columns of interambulacral plates occur and these are only a little

**Above** *Melonechinus multipora*, 7.5 cm in diameter, from the Carboniferous of Missouri.

larger than the ambulacral plates. The plates lack large tubercles and spines, giving a more or less smooth appearance to the test. The peristome is small with a simple lantern and no perignathic girdle.

When *Melonechinus* died, unless it was rapidly buried, it tended to collapse, with the plates becoming dissociated so that only individual plates or small fragments of the test are found as fossils. Careful examination of single plates is sometimes necessary to distinguish them from similarly dissociated crinoid calyx plates with which they may co-occur.

## *OVA*: EOCENE-RECENT, WORLDWIDE

An irregular echinoid with a heart-shaped outline, *Ova* has a rounded wedge-shaped profile and a rigid test up to 60 mm in length. The apical disc is small, posterior of centre, and has two gonopores. The anterior ambulacrum forms a deep groove – the anteal sulcus (see colour fig. 30) – which has undercut sides along its whole length and is non-petaloid. It leads to the peristome on the underside of the test. The other four ambulacra are petaloid, paired and slightly sunken into the test on the apical surface, flush adorally from the ambitus (the broadest part of the test). The anterior paired petals are strongly flexed and bowed and the posterior ones very short, bowed and barely sunken into the test. The pores are paired within the petals but outside of them, including the oral surface, they are single. There are two fascioles present – a peripetalous fasciole (around the area with the petals) which is sharply indented between the ends of the paired petals and continues over the anterior unpaired ambulacrum, and a lateroanal fasciole which extends from the anterior half of the peripetalous fasciole posteriorly and continues beneath the periproct. The interambulacra are very wide adapically and composed of very broad plates. On the oral surface they form a broad plastron which has two large plates opposite one another behind the labrum, an overlying lip-like modification of interambulacrum 5 which mostly hides the opening. The periproct is located on the flattened posterior of the test and is nearly the same size. The tubercles adapical to the ambitus are very small and very closely spaced, those of the oral surface are rather larger and more widely spaced.

Living schizasterids, of which *Ova* is one, excavate a burrow, building a funnel to the surface of the sediment and a drainage tunnel behind the burrow. They feed on small particles which are transferred to the mouth by the tube-feet.

**Above** *Ova canalifera*, 7 cm long, from the Pliocene of France; oral (upper fig.) and side (lower fig.) views.

**Above** *Temnocidaris (Stereocidaris) sceptrifera*, 5.5 cm in diameter, from the Cretaceous Chalk of Hertfordshire, England; apical (left), side (centre) and oral (right) views.

## TEMNOCIDARIS (STEREOCIDARIS): CRETACEOUS-RECENT, WORLDWIDE

This regular echinoid with a more or less globular test can be up to 50 mm in diameter. The interambulacra are each composed of two columns of large polygonal plates. These plates have a single large primary tubercle that is perforate and non-crenulate. The primary tubercles are surrounded by a ring of very much smaller tubercles. Over the remainder of the surface of the plate there are dense granules. There are between five and eight interambulacral plates per column. The ambulacra each have a pair of very much narrower, sinuous columns of simple plates, with uniserial pore pairs, one pair per plate, and tubercles. The apical disc is dicyclic (the genital plates form the margin of the periproct and the ocular plates occur in the angle between the adjacent genital plates outside this ring). The peristome is large, central and does not have any buccal notches (sometimes referred to as gill slits). The lantern is tall and has U-shaped teeth without keels when seen in section. The primary spines are large, solid, fusiform (i.e. long and tapering, with a fatter middle) and have many spinules over the surface so that they feel very rough when rubbed from the tip to the base. The ends may be splayed out in a rough cup-shape.

At the present day, the subgenus *Stereocidaris* is mostly found in deep-water and is omnivorous. It is relatively common in the Late Cretaceous Chalk.

## CRINOIDS

Crinoids have a long fossil record, the earliest examples coming from the Lower Ordovician, the Tremadocian-Arenigian crinoid *Aethocrinus*. A Cambrian fossil called *Echmatocrinus* was once thought to be the earliest crinoid but is now thought probably to be an octacoral. Crinoids were abundant and diverse during the Palaeozoic, with the major groups being established during the Ordovician. At the end of the Ordovician there was a decline in diversity until the middle of the Silurian when they recovered, only to decline once again in the late Devonian. Once more they became abundant and diverse in the lower Carboniferous, very often being the dominant or most abundant fauna in a reef, their stems and disassociated ossicles forming banks of what would later become crinoidal limestones. When crinoids die they disarticulate fairly rapidly, for within a few days of death the ligaments and soft tissues which hold the component skeletal elements together have decayed.

Where a crinoid has been preserved intact (or nearly so) it would have been buried rapidly and with sufficient sediment cover to avoid disarticulation of the component parts by water currents, scavengers or burrowing organisms.

Various groups of crinoids expanded through the Carboniferous until the later Carboniferous and Permian when a gradual decline set in. At the end of the Palaeozoic very few crinoid groups were left, and the great diversity of the Palaeozoic forms was over, and never recovered, with but one lineage surviving into the Triassic. This diversified into other groups during the Triassic and into the rest of the Mesozoic and beyond, with stalked crinoids being overtaken by diversifying comatulids which now dominate modern crinoid faunas.

Most of the fossil forms may be generally characterised by having a stem and a calyx with arms. They lived attached to the sea floor, sometimes by means of a well-developed holdfast; a weakly flexible stem of many disc-shaped ossicles developed from this. Some of these ossicles may bear slender, flexible cirri, which in modern, stemmed crinoids provide an alternative means of attachment to the sea floor. At the top of the stem is the calyx containing the gut and other soft body-parts, and the mouth, usually covered by a plated or leathery tegmen (the part of the calyx above the origin of the free arms forming the roof of the cup or calyx) and facing away from the sea floor. There is often a well-developed plated anal tube rising from the tegmen, and this may be a small rounded protuberance or a long tube. There are several food-gathering arms, which in some crinoids may bear fine, finger-like side branches termed pinnules, with food grooves in the pinnules and arms leading to the mouth.

Others crinoids, notably the modern feather stars (see colour fig. 34), are free living and drift or swim weakly and attach themselves to objects by means of cirri. They have stems in their very early development but quickly become detached. Some Palaeozoic crinoids may also have detached and lost their stems, and Jurassic

*Ailsacrinus* shows both states, with some still attached to stems and others with rounded calyces where the stem would have been.

Crinoids, of all the modern echinoderms, have a very much greater mass of skeleton in proportion to soft parts. Recent research has discovered that a modern crinoid can lose its calyx perhaps deliberately by autotomy, but the stem continues to exist. Evidence for this is seen in the proximal stem which has been rounded over. Other work has also discovered that fossil crinoids also had this ability. Nutrition is most probably obtained by direct diffusion through the stem epithelia, but reproduction by the stem is not possible, that function being reserved for the calyx which contained the gonads.

Crinoids are suspension feeders relying on their arms and pinnules to trap food particles by adopting several methods. Free-living comatulids actively seek favourable locations to feed, such as on the upper parts of reefs or attached to seaweed, and spread their arms to form a symmetrical, planar, filtration fan held into water currents. Alternatively, in more turbulent conditions they may adopt a radial feeding posture in which the arms are extended in many directions. A third method for deep dwelling comatulids is to collect material falling from above by arranging the arms to form a collecting bowl or funnel. Stemmed crinoids use a planar or a parabolic brachial filtration fan (see colour fig. 33), relying on currents for feeding, with the aboral side facing into the current. The arms and pinnules are used for food gathering, the tube-feet passing food particles into a central groove lined with mucous which is kept moving by cilia. Fossil crinoids are presumed to have fed in the same way as their modern descendants.

Predators do not seek out crinoids actively as food, perhaps because of a toxic substance in their food transferring mucus, so they may be considered to have few enemies. However, they do have parasites such as gastropods or polychaete worms which bore onto the stem and arms, forming cysts. Holes surrounded by cysts are often preserved in fossil crinoids and have been given the trace fossil name *Tremichnus*. The Silurian parasitic gastropod *Platyceras* lived close to the anal tube of its host crinoid and may also be preserved with the crinoid. Commensals include crustaceans and some ophiuroids.

Stemmed crinoids were very much more common in Palaeozoic times than subsequently and, although still living today, they are much rarer than the free-living species and are inhabitants of the deep sea. From the viewpoint of a zoologist, the free-living comatulids are typical crinoids, but for palaeontologists the stemmed forms are regarded as being more characteristic. There is good evidence that a transition between attached and free-living modes of life has occurred independently several times during the evolutionary history of crinoids. Some of the fossil free-living crinoids found in the Palaeozoic (e.g. *Myelodactylus*) and Mesozoic (e.g., *Ailsacrinus*) are not closely related to the modern comatulids, and had different lifestyles.

Crinoids are found worldwide from the lower Ordovician to Recent. Modern species, of which more than 600 have been described, live in a variety of marine environments, from the cold waters of great ocean depths and polar regions, to shallower warm tropical waters. Those living in colder waters generally have fewer and longer arms (up to 10) than those from warmer waters which often have more than 10 arms.

## *ACTINOCRINITES:* CARBONIFEROUS, EURASIA, NORTH AMERICA, NORTH AFRICA AND AUSTRALASIA

*Actinocrinites* has a calyx which is conical and elongated and a rigid, domed tegmen made of numerous small plates. The calyx is monocyclic, i.e. a basal circle of plates supports the radial circlet, with three low basal plates of equal size, and five hexagonal, elongated radial plates within which is a long pentagonal primanal plate (the lowest plate of plates called the 'anal series'). The plates usually have radial ribs. The plates of the tegmen may

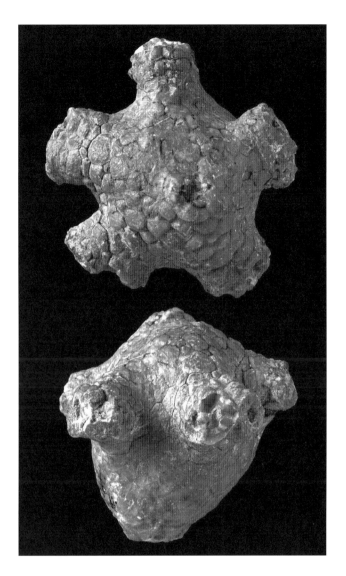

**Above** *Actinocrinites coplowensis*, 4.2 cm across the arms, from the Carboniferous of Lancashire, England; apical (upper fig.) and side (lower fig.) views.

have tubercles or spines, and an anal tube rises from it. Parts of the arms are incorporated into the calyx – the primibrachials (the first of the arm plates which are in contact with the radial plates) and the secundibrachials (two plates which branch from the primibrachial plates). There are also some additional plates included within the calyx. The arms are uniserial (one plate on top of another) when they are fixed, biserial (alternating plates forming two columns) when they become free, and are unbranched thereafter. The stem is circular and heteromorphic, i.e. with ossicles which differ in size and sculpture adjacent to one another. They are articulated by having interlocking crenellations (or 'symplectial articulation'). The stem, horizontal at this point, was attached to a substratum by the radices (cirri-like distal appendages).

In 1821, J. S. Miller erected the Class Crinoidea, and described *Actinocrinites* (as well as other genera) as a member of this class. Many species have since been assigned to *Actinocrinites*, but it is very likely that a lot of these are varieties of a few species. Specimens of *Actinocrinites* from the Permian of Timor are very variable and may represent extremes of form or even different taxa.

**Above** *Ailsacrinus prattii*, 3.4 cm wide, from the Jurassic of Bath, England, showing the underside of the calyx and arms.

## AILSACRINUS: JURASSIC OF EUROPE

In this genus the stem is very short (usually fewer than 10 columnals in length) and tapers away from the calyx, terminating in a rounded columnal. The columnals have symplectial articulation (interlocking crenellations) and the column may contain incomplete columnals that do not extend around the full circumference. The calyx is rounded or conical, with small basals and tiny accessory plates sometimes developed between the basal plates. The accessory plates may be tuberculate. Pinnules arise on alternate sides of the arms and are carried on every second or third brachial plate. They have ambulacral grooves, cover plates and elongated pinnulars (the component ossicles of the pinnule). Specialised oral pinnules with transverse grooves lie across the oral surface of the calyx and lack cover plates and ambulacral grooves.

*Ailsacrinus abbreviatus* from its type locality near Northleach in Gloucestershire, England, is found in crowded masses on bedding planes, with some 200 hundred individuals of this unattached crinoid contained in an area of a square metre. Specimens are usually well preserved, with their oral surfaces uppermost in life position. Some even preserve remains of mauve or purple pigmentation. Most are relatively complete, and it has been suggested that rapid influxes of sediment caused the death of dense populations that lived with their outstretched arms interlocked to form a kind of stable mat over the sea bed.

## APIOCRINITES: JURASSIC AND POSSIBLY CRETACEOUS OF NORTH AMERICA, EUROPE AND NORTH AFRICA

*Apiocrinites* has a smooth pear-shaped to globular calyx with five pentagonal basal plates and five radial plates which have an inverted pentagonal shape. The body cavity is large and broad. The large first primibrachials abut one another, with each arm branching once at the second primibrachials. The arms are uniserial, long, and have pinnulate secundibrachials (the two branches which arise from the primaxillary, which is the highest primibrachial plate of a branching crinoid arm). Apart from the region close to the calyx, the column has a smooth, circular outline with columnals of equal width and height articulating symplectially (interlocking crenellations). It has a narrow axial canal. Near the calyx the column increases progressively in width and there is no sharp distinction between column and calyx unlike other stemmed crinoids. At the base of the column there is a large holdfast made from extensive stereom overgrowths. The holdfasts of adjacent individuals may coalesce giving the false appearance of a single animal having multiple stems.

*Apiocrinites*, like *Actinocrinites*, was one of the original crinoid genera named by Miller in 1821. One of the famous localities for *Apiocrinites* is Bradford-on-Avon in Wiltshire, England, where the Middle Jurassic Forest Marble and overlying Bradford Clay outcrop. During life many of the crinoids were cemented to the top surface of the Forest Marble which was a hardened layer (hardground) on the seafloor. They were swamped and catastrophically buried by an influx of mud that formed the Bradford Clay.

**Above** *Apiocrinites elegans*, 9.2 cm tall, from the Jurassic Bradford Clay of Wiltshire, England.

## GLENOTREMITES: CRETACEOUS OF EUROPE

The column in *Glenotremites* is reduced to a single, flattened to concave bowl-shaped ossicle. This star-shaped plate bears numerous, irregularly arranged, large, circular to elliptical scars laterally which mark the attachment points of finger-like cirri. The cup is monocyclic, with elongate basals only visible as low triangles laterally. The radials are large and have broad facets, which articulate with the arms. The arms are uniserial, branch isotomously at the second primibrachial, and are pinnulate.

**Above** *Glenotremites aequimarginatus*, 9 mm in diameter, from the Lower Cretaceous of Kent, England, seen in side view.

*Glenotremites* is fairly typical of the most diverse group of extant crinoids, the comatulids. They are more or less free living and are able to walk or swim with their arms and can grip a substrate with their well-developed cirri. They are found in shallow water and are nocturnal, avoiding predators by concealing themselves among seaweeds and crevices in rocks or reefs.

## ISOCRINUS: TRIASSIC-RECENT, WORLDWIDE

*Isocrinus* has a low, wide, apparently monocyclic calyx, with five small basals at the surface of the calyx which project slightly to overhang the top of the stem. The calyx is actually 'cryptodicyclic', in that the infrabasals (a circlet of plates beneath the basals) are concealed. Five much larger radials are in contact laterally. The arms are uniserial and branch equally (isotomously) about four times, from the second primibrachial plate. Pinnules are also present, commencing at the second secundibrach (i.e. the second of the free brachial plates). The columnals are pentalobate or roundedly sub-pentagonal (star-shaped) and have a small central lumen (hole). There are five petal-shapes forming the articular surface. Articulation is symplectial (interlocking crenellations). Cirri arise in groups of five at regular and frequent intervals from nodes on the stem. The stem is attached to the substrate by a cirriferous runner – a part of the stem which grows horizontally along the substrate.

The columnals of *Isocrinus* are the commonest and most distinctive of post-Palaeozoic crinoid remains. It should be noted, however, that the name *Isocrinus* is used in a very broad sense for fossils as a form genus. These star-shaped plates can confuse the unwary into believing that they may have found a tiny starfish. The petaloid appearance of the symplectial articulation is common to other genera of Isocrinida as well as the Pentacrinitidae.

**Above** *Saccocoma baieri*, a free-living crinoid, 3.5 cm across, from the Jurassic of Solnhofen, Bavaria, showing the calyx and arms in a block of Lithographic Stone.

## *SACCOCOMA:* JURASSIC OF EUROPE, AND POSSIBLY NORTH AFRICA AND NORTH AMERICA

This is a minute stemless crinoid with a bud-shaped calyx, five small basals and a minute centrale plate (a central pentagonal calyx plate which occurs inside the basal circlet). The radial plates are larger than the basals, and there are five oral plates on the tegmen. The arms are uniserial and very slender, and branch equally once and then into unequal branches distally, producing alternating ramules (minor branches of the rami or arms). The component plates are grooved and were probably equipped during life with cilia to move food particles along the arm to the mouth. The primaxial (i.e. highest primibrach plate) and the first few secundibrachs also bear paired dish-shaped lateral wing-like structures, whilst the later secundibrachs, further away from the calyx, carry paired vertical projections. There are no attachment structures.

*Saccocoma tenella* (Goldfuss) is commonly preserved as a complete crinoid in the Jurassic Solnhofen Limestone of Bavaria in Germany, frequently occurring in large numbers on a bedding plane. It was probably planktonic; the branches of the arms provided an extensive filter, and the paired wing-like structures acting as baffles to help keep the crinoid floating in its preferred

**Above** The stalked crinoid *Isocrinus robustus*, 14 cm tall, from the Lower Jurassic of Gloucestershire, England, showing the calyx and arms.

orientation. The vertical baffles when moved laterally produced water currents. A rather limited swimming capability has been inferred.

## SCYPHOCRINITES: SILURIAN-DEVONIAN OF NORTH AMERICA, EURASIA, NORTH AFRICA

*Scyphocrinites* has a large calyx, a stem of circular cross-section, and a large attachment structure (see colour fig. 32). The calyx is monocyclic, with five pentagonal basal plates and five larger polygonal radial plates, and expands or contracts towards the arm bases. The interradial areas are formed mostly of fixed pinnulars (plates forming part of the aboral skeleton of pinnules) and between ten and twenty or more fixed secundibrachs. The interradial areas can be slightly inflated or depressed, and may contract or expand towards the bases of the arms. The fixed arms are uniserial, the free arms are either uniserial or biserial, made of short brachial plates. Arms are pinnulate and branch isotomously. There is a robust anal tube positioned sub-centrally on the tegmen. The stem is heteromorphic, i.e. columnals are dissimilar along the length of the stem. Columnal articulations are symplectial (interlocking crenellations), and there is a pentalobate or pentastellate (five rayed) axial canal. Stems may be very long, as much as 3 m. The bulbous attachment structure is shaped like an orange, contains many unequal sized internal chambers, and is composed of tiny irregular plates derived from numerous rootlets or radices.

The attachment structures of *Scyphocrinites* were once of unknown origin, being described and named *Lobolithus* and *Camarocrinus*, but eventually they were identified as parts of a crinoid. Specimens from the famous Moroccan locality of Erfoud show how the rest of the crinoid relates to the attachment structure. It is now interpreted as a floatation device, allowing the crinoid to have a pelagic lifestyle. *Scyphocrinites* can occur in large numbers in local accumulations, preserved with associated floats, suggesting that individuals were perhaps linked together during life. In this way, the species was able to attain a wider geographical distribution than for crinoids attached to the sea floor.

## OPHIUROIDS

Ophiuroids typically have a sharply defined circular to pentagonal central disc and five thin, very flexible arms from two to twenty times longer than the diameter of the disc (see colour fig. 35). The arms are branched in some species. The aboral surface of the disc is composed of many plates, some of them considerably larger than others and radial in position. The madreporite is interradial and usually oral in position. The oral surface of the disc has a central mouth and five pairs of slits called 'bursal slits' bordering the arms that are the external openings through which sperm and eggs are released. Some ophiuroids may also brood their young but show no other parental responsibility. The ophiuroid skeleton is composed of many ossicles of complex shape, some of them resembling tiny vertebrae, and named accordingly, though they are not the same structures as those of vertebrates, arranged in four series in the arms – two lateral, one oral and one aboral. The vertebrae are linked one section to another by ball and socket joints. There may be long or short spines on the lateral arm plates but there are no pedicellariae.

The long-spined ophiuroids feed on detritus and also extract small particles from the water with their long tube-feet, passing the entrapped material to the mouth. Some species burrow, leaving just parts of the arms exposed to catch food particles, while others live on the surface, feeding on detritus and micro-organisms by catching them on sticky mucus that is secreted by glands along the arms and carried by ciliary action to the mouth. Because ophiuroids do not have an anus they have to feed on high-grade food which does not result in a lot of waste.

The short-spined ophiuroids are predators and are carnivorous, have short tube-feet and have very mobile arms, catching small worms and crustaceans by means of the flexible arm tips. Food is passed to tube-feet to be directed to the mouth, or by bending the arm to the mouth.

By the standard of echinoderms, ophiuroids are able to move rapidly. Ophiuroids with branched arms are filter feeders, but others can also adopt this method. There is no anus so waste matter is ejected via the mouth. In order to minimise the necessity to eject waste material, high quality food is sought. After death the ossicles separate from each other, so that intact complete fossil ophiuroids inevitably point to the exceptionally favourable conditions of rapid burial, calm sea bed conditions and/or the absence of scavengers.

Ophiuroids have not always had the form that modern or younger fossil ones have. In the early Palaeozoic there was less of a distinction between the disc and the arms, and some early starfishes display characteristics of both ophiuroids and asteroids together in the same organism. Ophiuroids occur worldwide from the Lower Ordovician to Recent, and today may be found from shoreline to abyssal depths. About 2000 extant species are known.

## OPHIURA: MIOCENE-RECENT, WORLDWIDE

*Ophiura* is a medium-sized ophiuroid reaching up to 40 mm in arm length. The large circular disc of *Ophiura* is covered in imbricating platelets. There are no disc spines or granulations. Radial and oral shield plates are enlarged, and a large centrodorsal plate (one of the first formed plates which lies aborally at the centre of the disc) is also present. There is no madreporite. Bursal slits (or gill slits, access slits to the internal gill pouches) open in the disc on either side of each arm and are supported by genital bars. The arms taper gradually and are composed of fused ambulacral ossicles, or vertebrae, with a peg and socket articular surface between vertebrae. Lateral arm plates are wrapped around the arms and bear short, stubby spinelets pressed close against the plates. Dorsal and ventral arm plates are also present forming a continuous series along the arm. The radial water vessel

**Above** *Ophiura bartonensis* from the Eocene of Hampshire, England; several individuals have been 'dissected' by weathering. The discs are 9 mm in diameter.

(the soft-tissue tube that supplies water to the tube-feet and the ampullae – fluid-filled bulbs that extend the tube-feet when squeezed by muscles) is enclosed by the ambulacral plates. The internal jaw frame is protected by spinelets.

*Ophiura* is a scavenger, eating detritus or small items of prey. Accurate taxonomic assignment requires knowledge of the details of pore and spine arrangement in the oral area. This is not always possible with fossils and consequently the true geological range of *Ophiura* in the strict sense is largely unknown.

## PALAEOCOMA: JURASSIC, WORLDWIDE

The disc is well defined and circular, covered in fine imbricate platelets which may have granules. The radial plates are large and form a prominent part of the disc, with other plates rather small and inconspicuous. There is no madreporite. Bursal slits open in the disc on either side of each arm and are supported by genital bars. The oral opening is protected by spinelets. The arms are robust, taper gradually, and are composed of vertebrae with a peg and socket articulation. The lateral arm plates are wrapped around the arm and carry rudimentary spinelets against the plates. Dorsal and ventral arm plates usually form a continuous series along the arm. The radial water vessel is enclosed by the ambulacral plates.

*Palaeocoma* is a large ophiuroid with arms up to 10 cm long. Overall it is similar to *Ophiura* but has finer aboral disc plating and differs in other details. It was probably a generalist scavenger, eating detritus or capturing small prey.

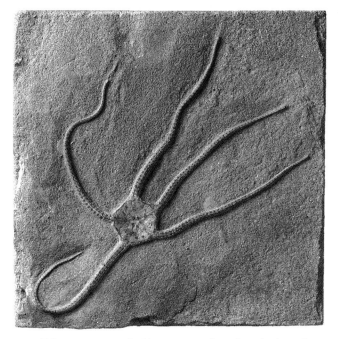

**Above** *Palaeocoma egertoni*, with arms 4.5 cm long, from the Jurassic of Dorset, England.

## ASTEROIDS

Asteroids are star shaped to cushion shaped (see colour fig. 36), often with five relatively broad arms, and lacking a distinct separation of the arms from the central disc. Depending on species, the number of arms may be fewer (four) or greater (fifteen or more) than five. The body is composed of a marginal framework of larger ossicles and covered elsewhere with a mosaic of smaller ones, some of them bearing small or robust moveable or fixed spines, granules and pedicellariae. Some of the spines may be attached to the lateral margins of the arms, others occur over the upper surface of the animal. The component ossicles fall into three groups according to where they are found: 'axial ossicles' which follow the line of the radial water vessels along the centre of each arm, and around the 'ring canal' (that part of the water vascular system which encircles the mouth and from which the radial canals extend), thus forming the oral or mouth frame; 'adaxial ossicles' which occur in series adjacent to the axial ossicles; and 'extraxial ossicles' which comprise all the rest. Marginals are enlarged ossicles around the outside of the starfish, with upper or superomarginals and lower or inferomarginals, usually placed one above the other. At the end of each arm there is a terminal ossicle which covers the ocular tentacle, a form of simple light sensitive eye. These terminals have characteristic shapes and can be useful when identifying different species. Muscles connecting the plates enable the starfish to move in various ways.

Starfish have a well-defined water vascular system, comprising a radial water vessel along the mid-line of each arm, with paired ampullae originating at regular intervals along its length. The ampullae control the extension and retraction of the tube-feet. The radial vessels are connected to the ring canal which is circum-oral. From the ring canal arises the stone canal (so called because it is frequently calcified), interradial in position, which connects the madreporite to the water vascular system. The madreporite is usually aboral in position, but in primitive starfish it is lateral. The anus when present is on the apical surface. Tube feet act as sensory organs, with other sensory cells scattered over the surface of the animal. The oral surface has a central mouth with grooves extending from it along the arms which carry the paired rows of tube-feet. The mouth frame is composed of the modified ends of the rows of axial ossicles.

Starfish subsist on a variety of food, including detritus and live micro-organisms, using ciliary action to transfer particles to the mouth. They may be carnivorous, swallowing whole large and active prey, such as hermit crabs and other crustaceans, sea-urchins and small fish.

**Above** '*Asterias*' *gaveyi*, the oral surface, with an arm stretch of 18 cm, from the Jurassic of Gloucestershire, England.

Some, like the modern 'Crown-of-Thorns' starfish, do great damage to coral reefs by grazing the polyps and leaving behind empty cups. Others, like *Asterias*, the common starfish, feed by turning the stomach inside-out through the mouth and onto the food source, allowing digestion to occur outside the body of the starfish. Mussels are consumed in this way after first opening the shells of these bivalves using the suckered tube-feet that clasp onto both valves of the shell. A sustained pull lasting a few minutes to hours results in a gap through which the stomach can be inserted. Digestion takes several hours, and when complete, the starfish moves on, leaving the empty shell. Scallops however, can detect the presence of a starfish by reacting to the touch of its tube-feet and flapping their valves to swim and escape. Starfish defend themselves against predators by means of toxic pedicellariae or by burrowing into the sediment.

Starfish, and also brittle-stars, are remarkable for their ability to regenerate parts which have been severed, including arms and the disc. Some starfish can even regenerate an entire body from just one arm and a part of the disc. This provides a means of asexual reproduction. For example, fishermen who cut up starfish from their catches and throw the pieces back into the sea may unwittingly actually increase their numbers. Starfish have lived worldwide from the lower Ordovician to Recent and nowadays exist from littoral to great depths.

*ASTERIAS:* JURASSIC-RECENT, WORLDWIDE
The disc is fairly large, with petaloid arms widest about a third of the distance from the centre of the body. Arm

length may exceed 100 mm. Ambulacral plates are elongated laterally, with large, biserial sutural gaps for the tube-feet to emerge from the ampullae. The adambulacral plates form a series parallel to the ambulacral plates, are elongate, blocky, and have spines. The rest of the body is covered by stellate ossicles with small sutural gaps for papillae – small, finger-like extensions of soft tissue used for gaseous exchange.

Modern *Asterias* is a voracious predator of bivalves using its powerful tube-feet to prise apart the two shells and inserting its stomach through the narrow gap to digest the prey.

### CALLIDERMA: CRETACEOUS-RECENT, EUROPE AND ATLANTIC

*Calliderma* is a large pentastellate starfish with arms reaching over 120 mm in length (see colour fig. 37). The body is framed by stout infra-marginal and supra-marginal ossicles, with the supramarginals meeting dorsally along the outer part of the arms. Marginal ossicles have well-defined nodes, but lack spines and granules. The ambulacral grooves of the oral surface are narrow and ambulacral ossicles are mostly hidden by overlying adambulacral ossicles. Small polygonal plates cover the rest of the disc, and aborally these are not differentiated into any series. There are no gaps for papillae.

This 'cushion-star' at the present day is a deep-sea animal feeding on detritus.

### NYMPHASTER: CRETACEOUS-RECENT, EUROPE AND ATLANTIC

A moderately large pentastellate starfish, the arms of *Nymphaster* reach over 60 mm in length. The pentagonal disc forms about a third of the radial length of the starfish and has a frame of infra-marginal and supra-marginal ossicles with an ornament of fine, dense pits. The marginal ossicles continue to form the clearly defined, narrow arms. The supramarginals meet along the perradius for most of the length of the arm. Within the marginal frame of the disc the aboral surface is

covered with small polygonal plates, and these are also present on the oral surface. There are no sutural gaps for papillae. On the adoral surface there is a narrow ambulacral groove along each arm, and the ambulacral ossicles are almost completely hidden by overlying small blocky adambulacral ossicles.

Represented by several species in the Late Cretaceous Chalk, modern examples of this starfish live in deep waters on soft substrates.

**Above** *Nymphaster coombii*, the oral surface, 9.8 cm from arm tip to opposite margin, from the Cretaceous, Lower Chalk of Sussex, England.

### TEICHASTER: CRETACEOUS-MIOCENE OF EUROPE

This large starfish has robust arms up to 100 mm long. The body margin has a double series of large, blocky infra-marginal and supra-marginal ossicles, with large, closely-packed circular pits for spines. The marginal ossicles continue outwards to form the lateral walls of the arms. The aboral surface of the body, towards the disc, and on the interradial areas of the oral surface, has

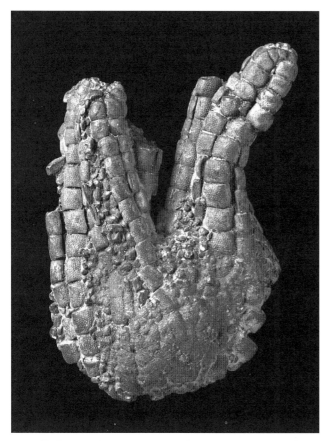

**Above** *Teichaster stokesii*, 3.6 cm across, from the Eocene London Clay of Kent, England; the specimen was folded over after death.

small, densely-packed polygonal plates. The ambulacral furrow is fairly broad and its ambulacral ossicles have wide interoscular gaps for the passage of the tube-feet. The adradial ossicles are small and blocky. In the dorsal parts of the arms, close to the main body of the starfish, a small number of polygonal plates separate the marginal plates of the arm, but further distally the supra-marginals meet towards the tips. The large, circular spine pits on the marginal ossicles are characteristic of this genus.

## HOLOTHURIANS

Holothurians are radially symmetrical and are mostly worm-like or cucumber-shaped animals, hence the common name sea cucumber. A ring of modified tube-feet surrounds the mouth, five double rowed ambulacra extend along the length of the body, and there is an anus at the opposite end of the body from the mouth. In mobile holothurians three of the ambulacral rows are ventral and two dorsal, but sedentary species have no preference for ventral or dorsal.

These echinoderms are soft-bodied with microscopic calcareous spicules, some of them resembling tiny anchors or wheels, embedded in the soft tissues. Consequently, holothurians are rarely found as 'body' macrofossils except in cases of exceptional preservation. Individual spicules can be collected by bulk-sampling and sieving of sediments. The shape of these spicules is very important in the classification of the animal, usually more so than the appearance of the whole body.

When holothurians move the mouth is at the leading end. Most have tube-feet along the ambulacra or scattered over the body, though in some species these are lacking. They may live on the surface of the sea bed (see colour fig. 38), be partly buried or even completely buried. Feeding involves the ring of oral tentacles which either catch food particles by entrapping them in sticky mucus and then pass them to the mouth in the curled tentacles, or scoop sediment into the mouth. Burrowers ingest the material through which they burrow.

The morphology of the oral tentacles indicates the mode of life. More active species are equipped with simple tentacles whereas sedentary holothurians living in burrows, rock crevices and so forth have multibranched tentacles which they spread out into the open where they can capture food in the flowing water.

The tough and leathery skins, which may sometimes be slippery, provide holothurians with defence against predators. In addition they may also have toxins in the body wall that can kill fish and other animals. A spectacular method of defence is the ejection through the anus of masses of white, sticky threads called Cuvierian tubules, which in certain species may also be toxic. These can entangle an attacker or confuse it, and in extreme cases the entire digestive tract may be ejected. The threads and the digestive tract are regenerated, assuming the animal survives the attack.

Holothurians are known from the Silurian to Recent, modern species being found worldwide in shallow to abyssal depths and in various substrates.

## *ACHISTRUM:* DEVONIAN-CRETACEOUS, PERHAPS PALEOCENE, OF NORTH AMERICA AND EUROPE

The body is cylindrical and up to 60 mm long. Microscopic anchor-shaped spicules were embedded in the skin. There are also ten ossicles forming a ring around the mouth. Tentacles surround the mouth and there is a straight gut.

A few hundred specimens of *Achistrum* are known from Mazon Creek in Illinois, a well-known fossil Lagerstätte. However, this holothurian was first described from isolated anchor-shaped spicules in sediment samples. It was probably a deposit feeder.

**Above** *Strobylothyone rogenti*, 2.9 cm long, from the Triassic Muschelkalk of Catalonia, Spain.

## *STROBYLOTHYONE:* TRIASSIC, SPAIN

The cylindrical body is up to 30 mm long. The body wall is composed of very thin, small, imbricating plates. There is a ring of ten oral ossicles. An anal cone of five plates is present at the opposite end. Pores for the passage of tube-feet are lacking.

*Strobylothyone* is known only from a Triassic Lagerstätte in Spain where several genera of exceptionally preserved holothurians occur.

## CYSTOIDS

Cystoids are a diverse extinct group of echinoderms recognized by having a theca of many plates with distinctive pores. They were equipped with food-gathering arms or brachioles, some had a stem which anchored them to the sea floor, some had a mere stem-like protrusion of the theca, and others were attached directly to the sea floor by an attachment structure. There are two major orders of cystoids – rhombiferans and diploporitids – distinguished by their thecal pores. In the rhombiferans, thecal pores are shared between adjacent plates, while in the diploporitids thecal pores are confined to each plate, usually but not always in pairs. These groups may be polyphyletic, i.e. evolved more

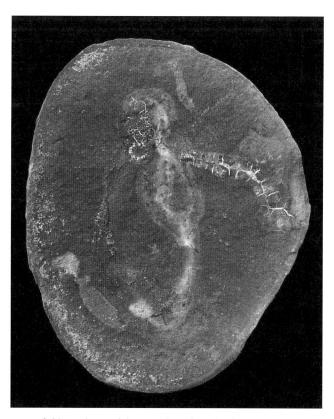

**Above** *Achistrum* in a nodule 7.4 cm long, from the Carboniferous of Illinois, showing the oral ring.

than once. The pattern of distribution of the pores over the plates of the theca is crucial to the classification of these echinoderms.

In diploporitids, the canals which open as 'diplopores' join to form a Y-shape beneath the thecal surface of the plate. Unpaired pores of diploporitids, however, do not form a Y-shape but instead meander beneath the surface, sometimes branching. Pores in some species of diploporitids are partly or wholly covered over by an epithecal layer. In rhombiferans, the often slit-like pores are usually arranged in a rhombic layout across plate boundaries termed the pectinirhomb. Sometimes only half of the rhomb is developed, the demirhomb. The canals may be covered over by epitheca. Each paired pore unit is shared by the two adjacent plates.

Thecae vary in shape and size, from relatively symmetrical to asymmetrical, and from those with numerous small plates to those with relatively few larger plates. Cystoid thecae are usually only a few centimetres in diameter, but can be much larger in some species, e.g. 10 cm in *Echinosphaerites*. The stem is variable in length and thickness. Some types of cystoids probably lost their stems altogether as juveniles, or perhaps never developed a stem at all.

There are several openings over the surface of the theca, interpreted as following the standard pattern of mouth, hydropore, gonopore and anus, a layout which is also found in carpoids and primitive chordates, amongst others. The anus may be formed into an anal pyramid as in some crinoids.

As with other echinoderms, cystoids have an ambulacral system, normally consisting of five grooves, but some genera have only two. Associated with these are biserial brachioles which are the appendages for food-gathering. The ambulacra transferred food from the brachioles to the mouth and both grooves were covered by two rows of small plates. Their brachioles were not as extensive or as robust as the arms of crinoids, and the openings in them were small so that only microscopic food particles could be ingested.

The major organs of the animal were mostly encased in the theca which served a double purpose of isolating them from the water and of protection, with the various apertures being protected by their own covering plates. Essential processes such as feeding, respiration and reproduction were carried out within these constraints. Feeding required that food particles gathered from water currents were passed along ciliated ambulacral grooves, perhaps by means of a mucus string, to the mouth, and respiration by means of podia from the diplopores (rather like the tube-feet of echinoids). Reproductive strategies are unknown but presumably followed the modern echinoderm method of shedding sperm and eggs into the sea where development occurred. Like other sessile marine organisms dispersal relied upon water currents.

Cystoids range from Ordovician to Devonian, reaching an acme of maximum diversity in the Middle Ordovician. They occur in shallow water shales, fine sandstones and limestones.

### *APIOCYSTITES:* SILURIAN-DEVONIAN OF EUROPE AND NORTH AMERICA

The column is divided into a number of morphologically distinct regions. It has a wide proxistele (the widest part of the whole column and adjacent to the theca, muscular in the proximal part), which tapers rapidly and has annular columnals. The dististele (furthest from the theca) is long and has cylindrical columnals. The theca is plum-shaped and round in cross-section and has four basal, five infralateral, five lateral, five radial and seven oral plates. There are three small rhomb plates which have external pores that open on the external surface as slits. The ambulacra are long and extend over the theca to the column. The periproct is small and is surrounded by three or four plates. The mouth is elongate and gently curved. The hydropore – an opening near the mouth – is semicircular, slit-like, close to the circular gonopore, and both are close to the mouth.

**Above** *Echinosphaerites aurantium*, side view, 2.4 cm in diameter, from the Ordovician of Russia.

**Above** *Apiocystites pentremitoides*, 2.6 cm long, from the Silurian, Wenlock Limestone of Dudley in Worcestershire, England.

The structure of the column indicates functional differences between the parts. The animal may well have been able to move in a similar fashion to that inferred for carpoids (see below). Rhombiferans such as *Apiocystites* have rhombic respiratory canals or rhombs, with half of each set occurring on two adjacent plates.

### *ECHINOSPHAERITES:* ORDOVICIAN OF NORTH AMERICA AND EURASIA
The theca is globular, rather smooth, and consists of 200 or more thin and smooth or finely striated plates, with narrow, compound fistulipores (thecal canals that have internal pores). The brachioles are branched and biserial, usually numbering three or four but occasionally only two. They arise from an expanded oral protuberance with a narrow neck surrounding the peristome. Secondary cover plates occur over the axis of the food

grooves, and these are small and arranged irregularly. The periproct has an anal pyramid and is present on the upper part of the theca and separated from the gonopore by several other plates. The column is much reduced and made up of single plates.

In North America and Scandinavia, *Echinosphaerites* occurs at distinct horizons and may form cystoid limestones. These cystoids, solid and filled with radiating sparry calcite, are known in folklore as 'crystal apples'. In the British Isles *Echinosphaerites* occurs as natural moulds in siliciclastic sedimentary rocks.

### *STAUROCYSTIS:* SILURIAN OF EUROPE
The theca is ovate or sub-ovate and composed of four basal, five infra-lateral, five lateral, five radial and seven oral plates. There are three disjunct pectinirhombs (dichopores that open as slits). When viewed from the oral end the theca has an approximately octagonal outline which is produced by the four raised ambulacra

**Above** *Staurocystis quadrifasciata*, 6.5 cm long, from the Silurian, Wenlock Limestone of Dudley in Worcestershire, England.

apical in position. The column is divided into a number of distinct parts, with the columnals being holomeric (i.e. each is composed of a single crystal component). Proximal columnals are wide, rapidly tapering and flanged, whereas the distal columnals furthest from the theca are long and cylindrical columnals. At the extreme distal end of the column there is an attachment structure of root-like branches.

The only species of this genus is from the Much Wenlock Limestone Formation of the English Midlands where it is the commonest cystoid found.

## BLASTOIDS

These small echinoderms have a stem, a calyx and arms. The stem is made up of thin disc-shaped columnals with a central hole and crenulations that interlock with those on adjacent columnals. The stem is slender, relatively short and was attached to a substrate with rootlets, like the holdfast of many crinoids.

The calyx contained the soft body parts of the animal and is pyramidal to globular in shape. The upper surface has five wide petaloid or narrower linear ambulacra, radial in position, extending from the apex down the calyx, from just more than halfway to beyond the basal elements. These bear many thin arms or brachioles used for food gathering in the living animal. Each ambulacrum consists of side plates separated by a lancet plate (an elongate, spear-shaped plate) with a median groove. The lancet plate extends from the tip of the ambulacrum to the mouth.

The calyx is composed of basal, radial and deltoid plates. The stem joins the calyx centrally in the basal circlet. The basals consist of three plates, two large and one small. The larger actually consist of two plates fused together. Some blastoids have all three plates fused together. Overlying and alternating with the basals are the radial plates which have a deep sinus extending from the apex downwards. This is occupied by the ambulacra. The deltoids are situated above the radials and are interradial in position. They are approximately

that are long and extend to the column. The brachioles are closely spaced, the peristome is presumably apical but is not visible, and the periproct is small, with an anal pyramid comprising three plates plus the corner of a fourth bordering it. The hydropore is dumb-bell shaped, and very close to a small circular gonopore. Both are triangular in shape, with the apex of the triangle bordering the central mouth. This part is referred to as the deltoid lip. In total there are 18 to 24 main plates in a calyx. Numerous tiny covering plates were also present but are not usually preserved.

At the apex of the calyx are several other openings. There are five or ten spiracles – circular or elongated openings adjacent to the mouth. One of these may also be combined with the anus and this is called the anispiracle. The anus may also occur as a separate opening between the posterior deltoids and limbs of adjacent radials. Within the calyx are thin-walled folded structures either side of the ambulacra, excavated into the adjacent deltoid and radial plates, and with a canal leading to the appropriate spiracle. These are the hydrospires and the hydrospire canals. Tiny openings next to the ambulacral margins are hydrospire pores. Water entered fissiculate blastoids via hydrospire slits to the hydrospires and evacuated from the spiracles or the summit plates. In spiraculate blastoids, water entered through the pores to the hydrospires and from there via the hydrospire canals to the spiracles. In blastoids which had neither pores nor spiracles, the water probably entered and left through the same hydrospire slits. The purpose of the entry and exit of water was for respiration.

Blastoids were most likely small-particle feeders as suggested by the structure of their brachioles, with food suspended in water taken, possibly by mucous strings, along brachial food grooves to the mouth at the summit of the calyx.

Blastoids are found worldwide and lived during the Silurian to the Permian, attaining their greatest diversity in the Carboniferous. They frequently occur in large numbers and may form limestones in a similar fashion to crinoids as in the Carboniferous of northern England and elsewhere. In the Permian of Timor they occur in impressive numbers, both in quantity and numbers of taxa. They may be found intermixed with other echinoderms and shelly debris of reefs and their associated structures.

## DELTOBLASTUS: PERMIAN OF AUSTRALASIA AND SOUTHEAST ASIA

The theca is ellipsoidal, with a concave base, and is a rounded polygonal in oral and basal views, with an overall five-rayed symmetry (see colour fig. 39). There are ten spiracles, an anal opening, three small basal plates and radials which are less than half the height of the theca and overlap the tall deltoids. The lancet plates are exposed and form petaloid ambulacra. There are two internal hydrospire folds on either side of each ambulacrum. The anus is close to the mouth.

*Deltoblastus* is especially abundant in Timor, Indonesia.

**Above** Oral view of *Ellipticoblastus ellipticus*, 1.9 cm in diameter, from the Carboniferous of Yorkshire, England.

## ELLIPTICOBLASTUS: CARBONIFEROUS OF THE BRITISH ISLES

The theca is high and almost globular, slightly concave basally, and is a rounded pentagon in oral and basal view. It has five spiracles around the mouth, one of which is a large anispiracle – the largest spiracle, posterior in position and which includes the anal

**Above** *Pentremites sulcatus*, 2.3 cm in diameter, from the Carboniferous of Illinois; side (left) and oral (right) views.

opening. There are three small basal plates which are hidden in lateral view; the radial plates are over half the height of the theca and the deltoids are overlapping and well defined. There is one superdeltoid plate – one of a group of small plates referred to as the anal deltoids, flanking the anal opening and bordering the mouth; two cryptodeltoid plates – a pair of several anal deltoids flanking opposite lateral margins of the anal opening; and one hypodeltoid plate – one of a group of anal deltoids present flanking the anal opening on the aboral side, away from the mouth. On either side of each ambulacrum there is one internal hydrospire fold.

### PENTREMITES: CARBONIFEROUS OF NORTH AND SOUTH AMERICA

The theca is club-like to sub-pyriform (nearly pear-shaped) in lateral outline, and polygonal with an overall five-rayed symmetry in oral and basal views. It tapers rapidly to a narrow base. There are five spiracles around the mouth, including a large anispiracle, the largest spiracle having the anal opening and being posterior in position. There are three low basals, the radials are moderately high and overlap the low deltoids. In the oral and anal areas there are many imbricate plates. There are three to seven internal hydrospire folds each side of an ambulacrum.

*Pentremites* is one of the commonest blastoids found in the Mississippian (Lower Carboniferous) of North America.

### EOCRINOIDS

Eocrinoid fossils normally consist of a theca, brachioles and a tapering stem. However, in some taxa the stem may be lacking or much reduced. They generally show the typical pentaradial symmetry of echinoderms, though some eocrinoids are flattened and as a result certain of the rays may not be developed.

The stems of eocrinoids vary in structure. They can be a hollow protrusion of the theca and composed of many irregular plates connected together loosely, such as that of *Gogia*. Alternatively they can be composed of numerous thin, cylindrical columnals and taper distally. The distal end of the stem may have a rounded point or may spread out to form a holdfast. There is an axial canal through the centre of the column, greater in diameter at the proximal end where it opens into the body of the theca.

The theca is composed of numerous plates and is irregularly spherical, pyriform, conical or cylindrical in shape. It is normally rigid but can be flexible, and the plates are imperforate. It has up to four openings: mouth, hydropore, gonopore and anus. Sometimes the hydropore and gonopore are combined to form a hydro-gonopore. Along the sutures of the plates there may also be pores – sutural pores – which presumably had soft tissues protruding from them. The purpose of the soft tissues is not known but was likely to be for gaseous exchange or chemoreception. Five food grooves lead to the mouth and these are shielded by cover plates.

The brachioles are simple structures arranged around the edge of the upper surface of the theca. They functioned in gathering food and transferring it to the mouth. Unlike the stem, they are neither hollow nor continuous with the body cavity of the theca. Food grooves in the brachioles connect with those of the theca.

Eocrinoids range from the lower Cambrian to the Silurian, and are known from Europe, USA, North Africa, Australia.

## CYMBIONITES: CAMBRIAN OF AUSTRALIA

This eocrinoid is small, rounded bowl-shaped, almost spherical, and is composed of five plates whose sutures are best seen from the inside of the bowl. The central cavity is conical and deep, the upper surface scalloped.

*Cymbionites* is known from the Middle Cambrian of Australia and is considered to be the basal circlet of an eocrinoid. It is usually found associated with disarticulated eocrinoid thecal plates. Previously, it was assigned to its own subphylum and class, and until more complete specimens are discovered, the exact nature of the fossil will remain unknown.

**Above** *Cymbionites craticula*, a slab 13 cm across, with several individuals, from the Cambrian of Queensland, Australia.

## RIDERSIA: CAMBRIAN OF AUSTRALIA

The high cup of *Ridersia* is conical to almost cylindrical. It has three circlets of plates: four basal plates, five

**Above** *Ridersia watsonae*, several individuals, the largest 5 cm long, from the Upper Cambrian of Queensland, Australia.

infralateral plates and five lateral plates. The oral surface is flat and has six large oral plates which are separated by five ambulacra. There is a hydropore, a gonopore and an anal pyramid interradially. At the ends of the ambulacra there are two upright biserial brachioles. The slender column is heteromorphic and composed of low columnals. It is tapered proximally and becomes cylindrical distally.

# EDRIOASTEROIDS

Edrioasteroids have the appearance of a small starfish clutching a many-plated cushion, and are small to medium in size. They were stemless, flexible, sessile animals without appendages, composed of a theca with five, well-defined ambulacral food-grooves radiating from a central mouth. There are two other openings, one of them is the periproct in the form of an anal pyramid, the other has been interpreted variously as a hydropore, gonopore or hydrogonopore. The interambulacral theca is composed of many polygonal plates of varying numbers which may imbricate or be scale-like. The shape of the theca varies from a slightly inflated disc to globular and sac-like. Many species are usually found with the theca collapsed. The ambulacra are confined to the adoral surface and are either curved or straight. Curved ambulacra may be clockwise (solar) or anticlockwise (contrasolar), though in some edrioasteroids different ambulacra curve in different directions. Ambulacra may reach the margin of the theca or there may be a rim of smaller interambulacral plates beyond them. The ambulacra enclose the food groove which is composed of uni- or biserial floor plates, and is covered by biserial cover plates.

The plates of the peristome are composed of ambulacral and interambulacral elements. The periproct is found on the adoral surface in the posterior interambulacrum, laterally in some species. The plates composing it may be in contact with those of the peristome and occupy much of the posterior interambulacrum, or they may be central, or close to the posterior margin. Two kinds of plates form the periproct, those which cover the opening and form an anal pyramid, and those which surround this structure.

The third aperture present is found behind the peristome on the posterior interambulacrum. It is assumed to be present in all edrioasteroids but is only observed in some. This aperture has been interpreted as a hydropore for water intake and equalization of pressure in the water vascular system, functioning like the madreporite of an echinoid. Alternatively, it may have been a gonopore for the passage of reproductive products, or possibly served both purposes.

Edrioasteroids occupied a variety of environments, including sandy, muddy and limey sea-beds, but most seem to have preferred the latter. Some species of edrioasteroids apparently lived upright in muds but most attached themselves to hard substrates such as shells, stones or exhumed concretions, often clustered together and oriented in parallel. Unlike most animals attached to hard substrates, edrioasteroids could move and reorientate themselves. Upper Ordovician deposits in the Cincinnati region of the USA are particularly well known for edrioasteroids. These are most often found fixed to shells of the brachiopod *Rafinesquina*. Juvenile edrioasteroids evidently colonized the edges of living brachiopods but with growth moved onto dead, overturned brachiopod shells, utilizing both exterior and interior surfaces of the host shells. Catastrophic influxes of muddy sediment resulting from storm activity buried the brachiopods with the edrioasteroids still in life position.

Edrioasteroids range from the Lower Cambrian to the Lower Carboniferous, and are especially common in the Ordovician of North America and Europe.

## *CARNEYELLA:* ORDOVICIAN OF NORTH AMERICA

The disc-shaped theca of *Carneyella* has a domed upper surface and a flat uncalcified lower surface. It is up to 20 mm in diameter and is usually found in a collapsed

**Above** *Carneyella pilea*, 1.2 cm in diameter and attached to a brachiopod shell, from the Upper Ordovician, of Kentucky.

state. There is a well-developed margin of plates which forms the thecal frame. The five ambulacra are slender and curved, four curving anti-clockwise and one clockwise. They extend to the margin and become parallel to it. The ambulacra are composed of uniserial floor plates which are not visible from the outside and each of which supports cover plate on either side. The cover plates are arranged biserially. The oral area has three large cover plates and a hydropore plate. The interambulacral areas are composed of weakly imbricating plates. There are no sutural pores between them. There is an anal pyramid in the centre of the posterior interambulacral area.

*Carneyella* lived attached to hard substrates on the sea floor, and is frequently found colonizing brachiopod shells. It was a suspension feeder using the ambulacral tube feet to capture organic material.

## ISOROPHUSELLA: ORDOVICIAN OF NORTH AMERICA

The theca of *Isorophusella* is up to 20 mm in diameter, discoidal in plan-view with a domed upper surface and a flat uncalcified lower surface. There is a well-developed marginal zone of plates which forms the frame of the theca. The ambulacra are slender, with four curving anti-clockwise and the fifth clockwise, reaching the marginal rim of plates and turning parallel to it. They are composed of uniserial flooring plates which are not visible outside, and biserial cover plates arranged in many series, with primary, secondary and sometimes tertiary series present. The oral cover plates are not well differentiated. The hydropore is covered by many small plates. The interambulacra are composed of weakly imbricating plates which have no sutural pores. There is an anal pyramid in the centre of the posterior interambulacral area.

Like *Carneyella*, *Isorophusella* typically lived attached to brachiopod shells and was a suspension feeder.

**Above** *Isorophusella incondita*, 1.5 cm in diameter, from the Middle Ordovician of Ottawa, Canada.

## RHENOPYRGUS: ORDOVICIAN-SILURIAN OF EUROPE AND AUSTRALIA

The theca is turret-shaped and has a small apical oral area. The stem is long, pedunculate and composed of many imbricating plates in an approximately organised series of spirals or columns. The ambulacra are short and straight, form prominent ridges and extend to the edge of the disc. They are composed of uniserial flooring plates which are not visible externally, and biserial cover plates, with each flooring plate supporting a cover plate on either side. In the centre of the theca are oral

**Above** *Rhenopyrgus grayae*, 2.1 cm long, from the Upper Ordovician, of Ayrshire, Scotland; natural external mould (left), latex cast of it (right).

cover plates and a hydropore plate. The interambulacra are composed of a few small imbricate plates which form a tesselated pavement. There are no sutural pores. An anal pyramid is situated in the centre of the posterior interambulacral zone.

*Rhenopyrgus* lived partially embedded within the sediment and was able to vary its length by inflation and contraction of the stem. It may therefore have been able to withdraw into the sediment when threatened.

## HELICOPLACOIDS

Along with the edrioasteroids, the helicoplacoids are the first group of definite echinoderms to appear in the fossil record. However, their relationship to other groups of echinoderms is uncertain. They are distinctly spirally symmetrical rather than radially or pentamerically symmetrical. The fossils are usually found collapsed or flattened.

They were small animals (1-5 cm) with tiny, spirally arranged, un-sutured plates forming a flexible, spindle-shaped test, somewhat tear-drop shaped, and a lateral mouth situated adapically of the mid-height of the test. The test had the ability to vary in shape from pyriform when retracted to fusiform when expanded, imbrication of the plates allowing this to occur. Areas interpreted as ambulacra and interambulacra are present. The interambulacra are three columns of plates which fold over one another when the test was retracted so that the central column was on the outside and the outer columns lay beneath it. When expanded the central column forms a ridge, resembling the ridge tiles of a house roof with roof tiles on either side. The lower third of the body has no ambulacra and the interambulacra are arranged vertically rather than spirally.

Three ambulacra are present, two spiralling adapically and a third spiralling towards the base. The spiral arrangement ensures the maximum exposure to the water and hence to food particles. No pores have been identified in these fossils. The presence of a water vascular system has been inferred.

Current research suggests that helicoplacoids were suspension feeders that lived in a vertical orientation, with the pointed end of the body buried in fine-grained sediment. Very occasionally, fossil tests are found preserved in this attitude. Because of their shape and lack of support or anchoring structures, the point loading would have caused the animal to sink into the mud, but as this does not seem to have happened the implication is that the substrate was firm enough to prevent sinking. Aggregated populations of helicoplacoids appear to have been swamped by rapid influxes of mud, and are found grouped together on bedding planes, suggesting that they were gregarious. They are present in several facies of the Poleta Formation in California, though most commonly in the shales, suggesting that they were tolerant of a variety of environmental conditions. That

**Above** *Helicoplacus* sp., 5 cm long, from the Lower Cambrian of California; natural external mould (left), latex cast from it (right).

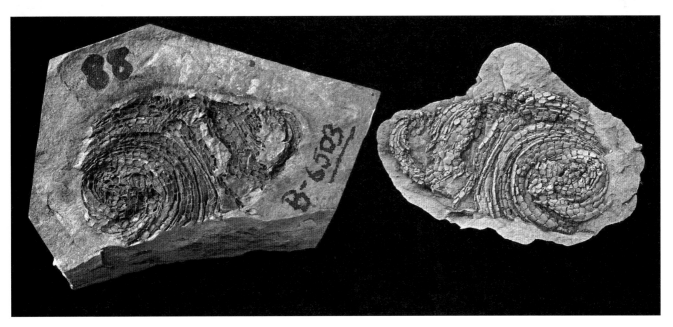

the most common facies are the shales suggests that elsewhere the preservation potential was lower, rather than the conditions being unsuitable.

Because the plates of the test were undoubtedly held together by soft tissue, where sudden burial did not occur after death, the skeletal elements would have disassociated quickly. Specimens are recorded with disrupted plate arrangement showing that the animal had died and begun to decompose just before burial. Further delay would have resulted in wider dissemination of the plates. Increasing bioturbation during the Cambrian was probably responsible for the extinction of these fascinating animals.

Helicoplacoids lived during the Early Cambrian of California, North America.

### *HELICOPLACUS:* EARLY CAMBRIAN OF CALIFORNIA, NORTH AMERICA.

The test is a prolate spheroid (i.e. 'cigar-shaped'), wider in the middle and tapering at each end, and was flexible during life. The component plates are small, very numerous and arranged spirally. The mouth is situated at one side, adapical to the mid-height of the test. There are three ambulacra present, one spirals towards the base and two spiral adapically.

## OPHIOCISTIOIDS

This small group of free-living pentaradiate echinoderms has the body covered with plates and hollow tubular structures interpreted as large tube-feet. The living animals are thought to have moved around on their oral surface, with the aboral surface uppermost. Their outlines varied from lobate, oval or roundly pentagonal.

The oral surface contains a central mouth or peristome equipped with five jaws, each made of two components, and the whole structure is covered by a buccal membrane. Radiating from the buccal area are five broad ambulacra which terminate at the outer margin of the test. The ambulacra are composed of two lateral columns and one mid-line column of plates. The plates of the middle column have two rows of circular pores interpreted as tube-feet pores. There are five much narrower interambulacra and a madreporite. Aboral to the madreporite is another opening presumed to be the hydropore, gonopore or gonohydropore.

The aboral surface is composed of many irregular, thin plates which may be smooth or ornamented by tiny surface granules. There is no apparent pattern in the arrangement of plates and no complex apical system. The centre of the test has a single plate referred to as the dorsocentral. The periproct is enclosed in an anal pyramid which can be marginal or just adjacent to the dorsocentral.

Ophiocistioids range from Lower Ordovician to Carboniferous, and have been found in Europe and North America. They are assumed to have been mobile, some more so than others, because of the lack of an anchoring stem, lack of grooved ambulacra and the presence of large robust tube-feet.

### *EUCLADIA:* SILURIAN OF EUROPE

The test of *Eucladia* is approximately pentalobate. The ambulacra are composed of seven or more perradial and adradial columns of plates, and six pairs of alternating podia and an unpaired terminal tentacle.

**Above** *Eucladia johnsoni*, 13.2 cm across, from the Upper Silurian of the English Midlands.

The most proximal pair of podia are smaller in size and are situated in the peristomial region. The others increase in size towards the margin of the test. The interambulacra have five or six plates in a single column each. The madreporite is well developed and has branched grooves. It is situated in the adoral interray which also has several gonopores. The peristome and buccal apparatus is surrounded by twenty plates.

## CYCLOCYSTOIDS

These strange fossils have a wheel-shaped theca which consists of a marginal ring of numerous, small, imbricating plates surrounding a submarginal ring of many thick, wedge-shaped plates forming rigid sides to the theca. Oral and aboral discs cover the area contained by the submarginal ring, and were weakly calcified. These discs are normally found collapsed in fossils but were probably flexible during life and contained the soft body parts of the animal. Faceted openings on the oral surface of the submarginal plates lead to internal ambulacra which join near the centre of the disc. The oral disc was probably more flexible than the aboral disc and may have been slightly inflated above the general level of the theca. The animal lived attached to a surface by its flat aboral side and did not possess a stem.

The geological range of cyclocystoids extends from the Middle Ordovician to the Middle Devonian. Specimens have been found in North America and northern Europe in a variety of different rock types, suggesting that species inhabited a variety of depths and environments.

### *ACTINODISCUS:* ORDOVICIAN OF BRITAIN

The theca of *Actinodiscus* is wheel shaped and consists of a ring of blocky marginal plates, radially elongate, some 40-45 in number and each with two cup-shaped depressions on the outer side. There is a peripheral skirt of small, radially-elongate platelets, reducing in size distally. The central zone has radially arranged plates in six-fold symmetry. These are stellate, uniserial and branch dichotomously. The positions of the mouth, hydropore, gonopore and anus are uncertain.

**Below** *Actinodiscus wrighti*, 1.6 cm in diameter, from the Upper Ordovician of Ayrshire, Scotland; natural external mould (left), latex cast of it (right).

**Above** *Polytryphocycloides davisii*, 2.5 cm in diameter, from the Silurian of Ayrshire, Scotland; natural external mould with latex cast overlying.

## *POLYTRYPHOCYCLOIDES:* ORDOVICIAN-SILURIAN OF NORTH AMERICA AND EUROPE

The wheel-shaped theca has a ring of 40-60 blocky marginal plates or ossicles, usually with two cupules but there are also single cupules regularly spaced. The marginal ossicles are in contact distally and the inner part of the ring of ossicles forms a raised, radially elongated crest. There is a peripheral skirt of small, radially elongated platelets which become smaller distally. The central zone shows four-fold symmetry and comprises stellate ossicles radiating from the centre, arranged uniserially and branching dichotomously. The positions of the mouth, hydropore, gonopore and anus are uncertain.

## CARPOIDS

These strange extinct animals, which include the mitrates, cornutes and solutes, are conspicuously asymmetrical and consist of a theca (or head) and a stele (or tail). The thecal skeleton usually consists of robust marginal plates and either a few large or many small plates covering the rest of the dorsal and ventral surfaces. There is a mouth at one end of the theca and this is surrounded by small plates. One or two appendages, whose function has been much debated, are attached to the theca. These may have acted as sediment stirrers, directional controllers during movement, or for feeding. The plates of the dorsal and ventral surfaces may have been flexible in life, with the smaller plates being attached by an integument.

A flexible stele composed of many plates extends from one end of the theca. This was used in locomotion, permitting the animal to drag itself backwards through or over the sediment. The stele is thought to have been raised above the level of the head and thrust downwards into the sediment, using it as a fulcrum to drag the theca along. Lateral movement was probably controlled by the anterior appendages to affect the pitch and yaw.

Carpoids may have been deposit feeders, employing the appendages to stir up the sediment during movement, ingesting sediment through the mouth and passing it to the pharynx where food particles were extracted. Animals possibly could 'cough' to expel waste sediment, in much the same way as modern *Amphioxus* reverses ciliary action in its pharynx. Openings in the theca interpreted as gill-slits are very obvious in genera such as *Cothurnocystis* but obscure in others (e.g. *Balanocystites*) unless the internal structures can be observed.

The affinities of the carpoids are controversial, with some scientists regarding them as echinoderms (the 'traditional' view), while others interpret them as ancestral chordates. Carpoid skeletons are of calcite stereom, as in the echinoderms, but lack the radial or bilateral symmetry typical of this phylum. Another character supporting a chordate relationship is the left-side openings (gill-slits), as in primitive larval stages of modern chordates. Some modes of preservation allow observation of a very complex series of structures, interpreted as a well-developed nervous system. Recent discoveries of fossil trails with the associated animal have corroborated deductions about the way that carpoids moved using their tails.

Carpoids range from the Middle Cambrian to the Carboniferous and have been found worldwide.

## *CASTERICYSTIS*: CAMBRIAN OF NORTH AMERICA.

The theca of *Castericystis* is ovoid in shape and composed of many plates. There is a moderately long single appendage anteriorly which has prominent cover plates. The anal pyramid is present to the right posterior of the theca. The stele is long and robust, proximally nearly as thick as the theca and composed of many small plates, followed by a shorter, more organised plated central section and tapering to a serially plated hind stele with six or seven ventral spines.

Juveniles of these have been found attached by the distal tips of their steles to the steles of adults.

**Above** *Castericystis vali*, 7 cm long with juvenile 0.8 cm long attached at the bottom of the stele, from the Middle Cambrian of Utah.

## *COTHURNOCYSTIS*: ORDOVICIAN OF EUROPE AND POSSIBLY NORTH AMERICA

The theca has the shape of a medieval boot, framed with marginal plates and covered by a flexible, plated, dorsal and ventral integument, attached to the frame. Prominent slits are present on the left side. There is a median strut on the ventral surface. The mouth has two oral appendages of unequal length and which were moveable. A third fixed appendage extends from the

left anterior marginal plate. The stele, flexible in life, is long and divided into three parts. The proximal part is made of four rows of imbricated plates, the mid part is short, and the hind part is slender and has ventral ossicles with biserial dorsal plates.

A large number of specimens of this peculiar fossil were amassed from Girvan in Ayrshire (Scotland) by the famous amateur collector Mrs Robert Gray. The generic name of the animal derives from the shape of the theca, *cothurnos* being Greek for a boot.

## *PLACOCYSTITES:* SILURIAN OF EUROPE

The theca is elongated and more or less rectangular, almost bilaterally symmetrical. There are relatively few plates, with the horizontal V-shaped marginal plates being the largest and most robust. Other large plates are present on the dorsal surface, while the ventral surface has slightly smaller plates. The surfaces of the thecal plates are ornamented by numerous transverse ridges. They are steep anteriorly and shallow-angled posteriorly. There are four small oral plates adjacent to the mouth,

**Above** *Cothurnocystis elizae*, 2.7 cm wide, from the Upper Ordovician of Ayrshire, Scotland; natural external mould (left), latex cast of it (right).

and at the corners of these there are two long, moveable appendages of carving-knife shaped cross-section. The fore-stele is short, robust and composed of four columns of plates arranged in circlets and imbricating dorsally. The mid-stele is short and tapering towards the hind stele which is slender and elongated. The dorsal surface of the mid- and hind-stele are noticeably blade-like.

*Placocystites* is well known particularly from the famous Wenlock Limestone locality at Wren's Nest, Dudley in Worcestershire, England. The oral appendages of *Placocystites* would have given it greater stability on the sea floor by increasing its effective area. They may also have moved from side to side, stirring up sediments to be ingested for food. The ratchet-shape of the ornamenting ridges of the plates would make movement in any direction other than backwards less efficient. This backwards movement was probably accomplished by the tail being thrust into the sediment and the head hauled backwards. At the end of the stroke the tail would be pulled out, extended and thrust in again and the process repeated.

**Above** Ventral view of *Placocystites forbesianus*, 2 cm across, from the Silurian of Worcestershire, England.

## *RHENOCYSTIS:* DEVONIAN OF EUROPE

The theca is more or less rectangular and slightly elongate, nearly symmetrical. There are relatively few plates, with fewer larger ones dorsally and more smaller ones ventrally. Oral plates occur as a grill-like row adjacent to the mouth. There are two oral appendages. The stele comprises a short, robust fore-stele, composed of circlets of four columns of plates, that tapers towards the very short mid-stele. The hind-stele is blade-like.

*Rhenocystis*, one of the last carpoids, is known from the Devonian of Bundenbach in Germany. Associated with the body fossils are trails seemingly made by the animal moving through the sediment. It is evident from these trails that *Rhenocystis* moved backwards, pulled by the tail, with the convex ventral surface downwards and the oral appendages wagging sideways.

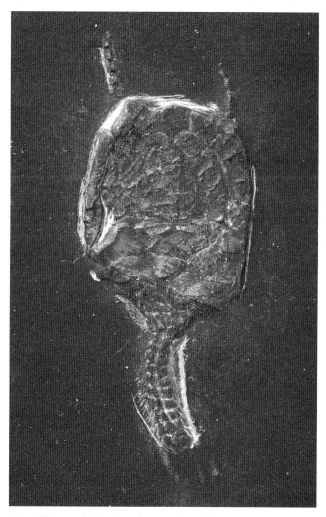

**Above** *Rhenocystis latipedunculata*, 3 cm long, from the Lower Devonian Hunsrück Shale of Bundenbach in Germany.

# Further information

Boardman, R.S., Cheetham, A.H. and Rowell, A.J. (eds) 1987. *Fossil Invertebrates*. Blackwell, Oxford.

Bottjer, D.J., Etter, W., Hagadorn, J.W. and Tang, C.M. 2002. *Exceptional Fossil Preservation*. Columbia University Press, New York.

Briggs, D.E.G. and Crowther, P.R. (eds) 2001. *Paleobiology II*. Blackwell, Oxford.

Brusca, R.C. and Brusca, G.J. 2003. *Invertebrates*. Second Edition. Sinauer Associates, Sunderland, Massachusetts.

Clarkson, E.N.K. 1998. *Invertebrate Palaeontology and Evolution*. Fourth Edition. Blackwell, Oxford.

Fortey, R.A. 2002. *Fossils: The Key to the Past*. Third Edition. The Natural History Museum, London.

Milsom, C. and Rigby, S. 2004. *Fossils at a Glance*. Blackwell, Oxford.

Monks, N. and Palmer, P. 2002. *Ammonites*. The Natural History Museum, London.

Murray, J.W. (ed.) 1985. *Atlas of Invertebrate Macrofossils*. Longman, Harlow.

Nielsen, C. 1995. *Animal Evolution*. Oxford University Press, Oxford.

Ross, A. 1998. *Amber*. The Natural History Museum, London.

Sheldon, P. 2001. *Fossils and the History of Life*. The Open University, Milton Keynes.

Stanley, S.M. 1993. *Exploring Earth and Life through Time*. Freeman, New York.

Taylor, P.D. 1990. *Fossil*. Eyewitness Guide. Dorling Kindersley, London.

Taylor, P.D. 2004. *Extinctions in the History of Life*. Cambridge University Press, Cambridge.

Tudge, C. 2000. *The Variety of Life*. Oxford University Press, Oxford.

Walker, C. and Ward, D. 1992. *Fossils*. Dorling Kindersley, London.

# Websites

### WEBSITES OF PROFESSIONAL PALAEONTOLOGICAL SOCIETIES

Palaeontological Association:
http://palass.org/index.html

Paleontological Society:
http://www.paleosoc.org/

Palaeontographical Society:
http://www.nhm.ac.uk/hosted_sites/palsoc/

### WEBSITES CONTAINING USEFUL INFORMATION ABOUT FOSSIL INVERTEBRATES

Fossil Folklore website:
http://www.nhm.ac.uk/fossilfolklore/

Museum of Paleontology, University of California, Berkeley: http://www.ucmp.berkeley.edu/

Paleontological Museum, University of Oslo:
http://www.toyen.uio.no/palmus/english.htm

Peabody Museum of Natural History:
http://www.nhm.ac.uk/museum/vr/index.html

Phil Bock's bryozoan website
http://www.civgeo.rmit.edu.au/bryozoa/default.html

Sam Gon III's trilobite website:
http://www.aloha.net/%7Esmgon/ordersoftrilobites.htm

The Echinoid Directory:
http://www.nhm.ac.uk/palaeontology/echinoids/

Virtual Wonders at the Natural History Museum, London:
http://www.nhm.ac.uk/museum/vr/index.html

# Index

# Picture acknowledgments

p.9, p.44 (l) p.69, p.106 (tl), p.111 (bl) © Paul Taylor

p.21 © NHM/Mike Eaton

p.56 © Peter Crowther

p.57 © Anton Kearsley

p.132 (l) © Steve Trewick

p.159 Redrawn by Mike Eaton from original by Annemarie Plint by permission of the Palaeontological Association

p.97 (l) © Masa Ushioda/ImageQuestMarine.com

p.97 (r), p.108 (tr) © Peter Parks/ImageQuestMarine.com

p.99, p.101 (b) © George Mitchell

p.106 (b) © Phillip Colla/Oceanlight.com

p.107 (br) © Roger Steene/ImageQuestMarine.com

p.109 (tl), p.109 (b), p.110 (br), p.111 (t), p.112 (t) © Kåre Telnes

p.110 (bl) © Charles G. Messing


All other images are copyright of the Natural History Museum, London. For copies of these and other images, please view the online Picture Library at www.nhm.ac.uk/piclib, or contact them directly at the Natural History Museum.

Every effort has been made to contact and accurately credit all copyright holders. If we have been unsuccessful, we apologise and welcome corrections for future editions or reprints.